Intro
Non-

Other titles in this series

Linear Algebra
R B J T Allenby

Mathematical Modelling
J Berry and K Houston

Discrete Mathematics
A Chetwynd and P Diggle

Particle Mechanics
C Collinson and T Roper

Ordinary Differential Equations
W Cox

Vectors in 2 or 3 Dimensions
A E Hirst

Numbers, Sequences and Series
K E Hirst

Analysis
P E Kopp

Groups
C R Jordan and D A Jordan

Statistics
A Mayer and A M Sykes

Probability
J H McColl

Calculus and ODEs
D Pearson

In preparation

Vector Calculus
W Cox

Modular Mathematics Series

Introduction to Non-Linear Systems

J Berry

Centre for Teaching Mathematics
University of Plymouth

A member of the Hodder Headline Group
LONDON • SYDNEY • AUCKLAND

First published in Great Britain 1996 by Arnold,
a member of the Hodder Headline Group,
338 Euston Road, London NW1 3BH

British Library Cataloguing in Publication Data
A catalogue record for this book is available from the British Library

ISBN 0 340 67700 7

Printed and bound in Great Britain by
J W Arrowsmith Ltd, Bristol

Contents

Series Preface

This series is designed particularly, but not exclusively, for students reading degree programmes based on semester-long modules. Each text will cover the essential core of an area of mathematics and lay the foundation for further study in that area. Some texts may include more material than can be comfortably covered in a single module, the intention there being that the topics to be studied can be selected to meet the needs of the student. Historical contexts, real life situations, and linkages with other areas of mathematics and more advanced topics are included. Traditional worked examples and exercises are augmented by more open-ended exercises and tutorial problems suitable for group work or self-study. Where appropriate, the use of computer packages is encouraged. First level texts assume only the A-level core curriculum. Higher level texts will either provide progression in a particular area of mathematics or introduce areas not included at lower levels.

Professor Chris D. Collinson
Dr Johnston Anderson
Mr Peter Holmes

Preface

'A butterfly flaps its wings in South America and days later a devastating storm creates havoc in southern Britain with the destruction of trees, occurrence of floods and much housing damage.' This so-called, 'Butterfly Effect' is often used to illustrate the relatively new area of applied mathematics called **chaos theory**.

The weather is a vastly complicated system and forecasting the weather for more than a few hours is probably based on unreal optimism. During the 1960s there were regular medium range and long range (i.e. monthly) weather forecasts, most of which were hopelessly wrong, and now the Met. Office at best succeeds in predicting the type of weather for two or three days only.

Discoveries in mathematics (and science) often occur by accident and the discovery of chaos in the solution of the computer model of the weather by Edward Lorenz was certainly a surprise for meteorologists in the early 1960s. Lorenz wanted to examine the solution of the weather model for a long period of time. He had a solution over a shorter period and instead of going back to the beginning with the same initial conditions he took a shortcut and started with data part way through the initial solution. What Lorenz expected was the same solution as the second part of the initial solution and then an extension for a much longer period. To his surprise the predicted weather pattern had changed considerably. The new solution was completely different from the original solution. The model was the same, but the initial data for the second run had been entered to a lower degree of accuracy than that stored in the computer. The small rounding error showed that given a slightly different starting point for the computer model, the weather would unfold in a completely different way. Thus the notion of the Butterfly Effect.

A simple form of the differential equations studied by Lorenz are

$$\dot{x} = -ax + ay$$

$$\dot{y} = rx - y - xz$$

$$\dot{z} = -bz + xy$$

where a, b and r are constants. The curve in space representing the solution of this system seems to loop around itself forever, never visiting the same point twice and never escaping. This feature has the peculiar name, **a strange attractor**.

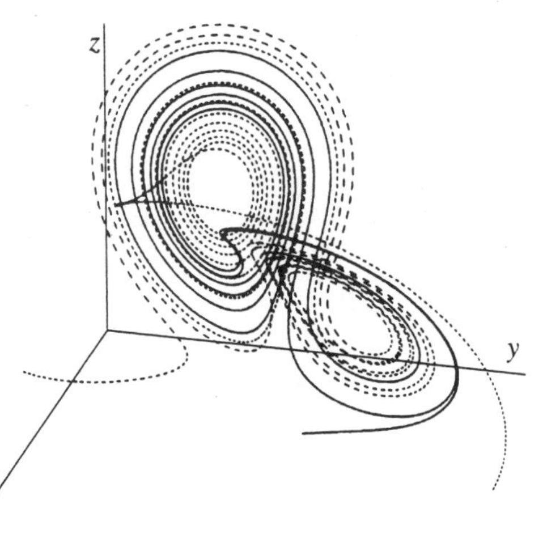

As we shall see it does not require a complicated model, such as those of atmospheric fluid dynamics, to demonstrate chaotic behaviour. The simple-looking quadratic recurrence relation

$$x_{n+1} = x_n^2 - c$$

shows quite unpredictable behaviour as the value of c changes.

For $c = 0.5$ and an initial value $x_0 = 0$, the recurrence relation converges to the point $x = -\frac{1}{2}(\sqrt{3} - 1)$. The figure shows a cobweb diagram of the iterations.

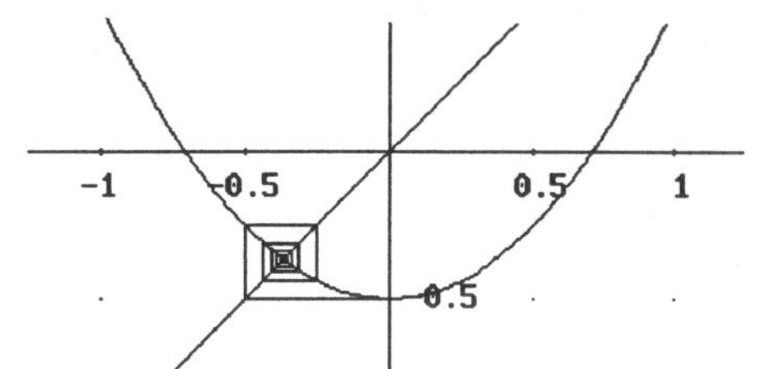

For $c = 0.8$ and initial value $x_0 = 0$, the recurrence relation is attracted towards an alternating pair of values: $-0.1(5+\sqrt{5})$, $-0.1(5-\sqrt{5})$. This is called a **periodic cycle**.

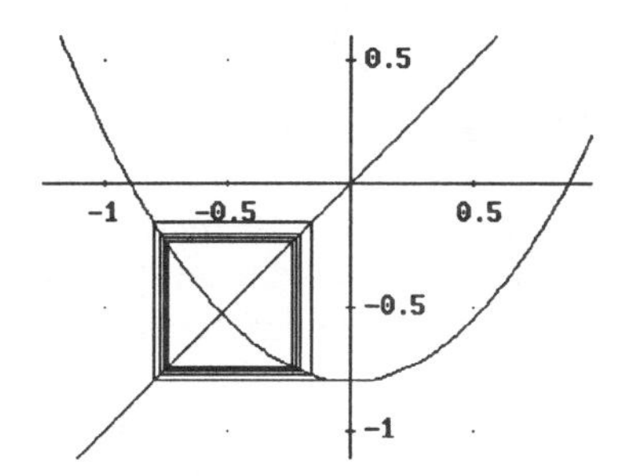

Increase c again, to say $c = 1.3$, and the following cycle occurs: 0.389019, –1.14866, 0.01943, –1.29962.

We can see that as c increases the behaviour of the recurrence relation takes on new features.

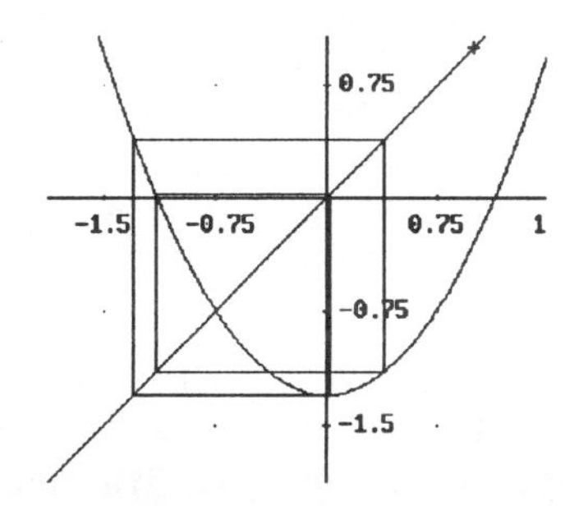

Related to the quadratic recurrence relation for real numbers is the beauty and fascination of Julia and Mandelbrot sets. These wonderful images are produced by investigating the complex system

$$z_{n+1} = z_n^2 - c$$

where z and c are complex numbers. This diagram is produced using the following simple algorithm:

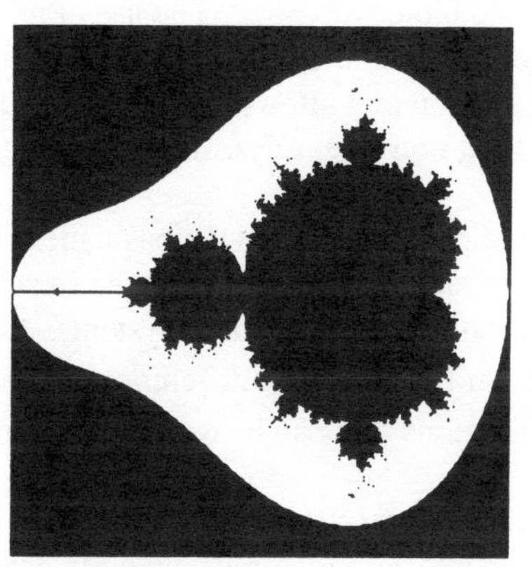

Choose a maximum number of iterations, n. For each value of c, iterate the recurrence relation n times starting with $z_0 = 0$. If the ith iteration is such that $|z_i| > 2$ for some $i < n$ then stop iterating and colour the point c in the complex plane white.

If $|z_i| \leq 2$ for all $i \leq n$ then colour the point c black. Repeat the algorithm for many values of c. The coloured version of this figure is produced by colouring the point c with different colours depending on pre-chosen ranges of values of n.

All of these ideas belong to the area of mathematics called **non-linear dynamical systems**. The term 'dynamical system' means an object or set of objects whose motion we are investigating. A simple pendulum swinging back and forth, the planets in the solar system, and the atmosphere near the earth's surface are all examples of dynamical systems. But systems do not have to be 'physical systems'. There are important applications of the theory to population modelling in biology, to the study of lead pollution of the environment, and to models of the stock market or the world economy to chemical reactions. The essential feature of a dynamical system is that its configuration or state is changing.

The laws governing the motion of the system will lead to differential equations or difference equations (i.e. recurrence relations) depending on whether the situation under investigation can be modelled as a continuous or a discrete system. When analysing dynamical systems the mathematician needs to solve these equations to predict the future behaviour of the system. Will it be sunny in July? Will the stock market rise or fall? Will the population of fish in the Irish Sea become extinct if more countries are allowed fishing rights? Clearly some dynamical systems are predictable. The sun will rise and set tomorrow (although you may not see it if it is cloudy); a child's swing in the park will swing back and forth today, next week and next year in the same way. But as Lorenz discovered, predicting next week's weather with any certainty is almost impossible.

In this book we provide an introduction to the basic mathematical concepts and techniques needed to describe and analyse non-linear systems. The approach taken is not that of the rigorous pure mathematician. There are many tasks suitable for the calculator or computer to encourage a feel for the subject. I believe that a first course in non-linear systems should demonstrate the essential features of what is going on in a concrete way. The abstraction to the underlying theories based on abstract analysis and algebra can then follow.

This book is an undergraduate text in dynamical systems aimed at students who have taken a first course in calculus. It has grown out of a course for second and third year students at the University of Plymouth. These students range from honours mathematics majors to biology majors. The use of technology such as Computer Algebra Systems or symbolic calculators (e.g. the Texas TI-92) reduces the demands of algebraic manipulation and allows students to focus in on the essential concepts and ideas in modelling non-linear dynamical systems.

The first chapter reviews the basic ideas of differential equations, matrix algebra and iteration methods that are assumed in subsequent chapters. Chapters 2 and 3 study first and second order continuous systems. Chapter 4 investigates discrete systems and in particular the quadratic difference equation $x_{n+1} = x_n^2 - c$. Chapter 5 is a collection of investigations that can be explored as more open-ended tasks.

In order to engage the reader actively in exploring the ideas in the book, each chapter contains tutorial problems which are designed to provide some of the essential theory to be understood. These problems are not just tasks that are similar to the worked examples. For the teacher using this book with students, you may find it convenient to stop at each tutorial problem and invite the students 'to have a go' at developing the next step in the argument themselves.

The availability of computer technology has allowed the analysis of non-linear dynamical systems to move forward rapidly during the last 30 years. For today's students, computer algebra systems are widely used and Chapter 4 makes use of the computer to explore the discrete quadratic system in some detail. The appendix contains the code used to generate many of the figures using the CAS DERIVE. Similar code can be used to explore the systems using other CAS such as Maple or Mathematica. My students are encouraged to enrich their learning of the theory of non-linear systems by many computer experiments.

Acknowledgements

It is a pleasure to thank Sharon Ward for preparing the manuscript as CRC and Mike Broughton for producing many of the figures. I am also indebted to the students of the University of Plymouth who have joined me in gaining an understanding of this fascinating topic. Together we have meandered through the mysteries of chaos!

Finally, I dedicate this book to my daughter, Elizabeth, whose gift of James Gleick's book Chaos on Father's Day 1995 resurrected the latent mathematician in me!

John Berry
Centre for Teaching Mathematics
University of Plymouth

1 •Review of Basic Ideas

The study of dynamical systems may be thought of as the analysis of differential equations for continuous models of the system or difference equations (i.e. recurrence relations) for discrete models of the system. Inevitably a course that investigates the motion of non-linear systems builds on some knowledge of calculus, differential equations and algebra. This first chapter is designed as a brief review of the basic ideas and methods needed for the rest of the book.

We begin with first order differential equations. In general the standard form of a first order differential equation is

$$\frac{dy}{dx} = m(x,y) \tag{1.1}$$

where x is an independent variable and y is the dependent variable. The fundamental problem is to find all functions $y(x)$ that satisfy the equation.

For a first order differential equation, the **general solution** will contain one arbitrary constant. Intuitively this makes sense because to solve an equation such as (1.1) we have to integrate once. If we have the value of y at a given value of x (called an **initial condition**) then the constant can be found and the solution is called a **particular solution**.

There are many different types and associated methods of solution of first order differential equations. In later chapters of this book you will only need to solve separable and linear first order equations analytically and you will need to be able to sketch solution curves from a knowledge of the direction field.

1.1 First order differential equations: Geometrical methods

The graphical approach to investigating differential equations via its direction field forms the basis of the methods in Chapters 2 and 3 for analysing non-linear systems, and is therefore of fundamental importance.

The basic idea is that the function $m(x,y)$ in equation (1.1) represents slopes to the solution curves of the differential equation. At different points of the x–y plane these slopes are represented by short line segments in the direction determined by $m(x,y)$. In this way a graphical picture of the solutions of the differential equation is built up. We can often sketch in solution curves from the general orientation of the line segments. The set of line segments is called **the direction field**.

Example 1

Draw the direction field of the differential equation

$$\frac{dy}{dx} = -\frac{x}{y}$$

Sketch in some solution curves for the differential equation. Suggest an equation for the general solution and verify that it satisfies the differential equation.

Solution

The calculation and drawing of the slope at chosen points of the x–y plane is a straightforward activity. For example, the following table and graph (Fig 1.1) show a few slopes for $m(x,y) = -\frac{x}{y}$.

x	y	$m(x,y)$
0.5	1.0	–0.5
1.0	1.0	–1.0
1.5	1.0	–1.5
2.0	1.0	–2.0
1.0	0.0	∞
1.0	–0.5	2.0
1.0	–1.0	1.0

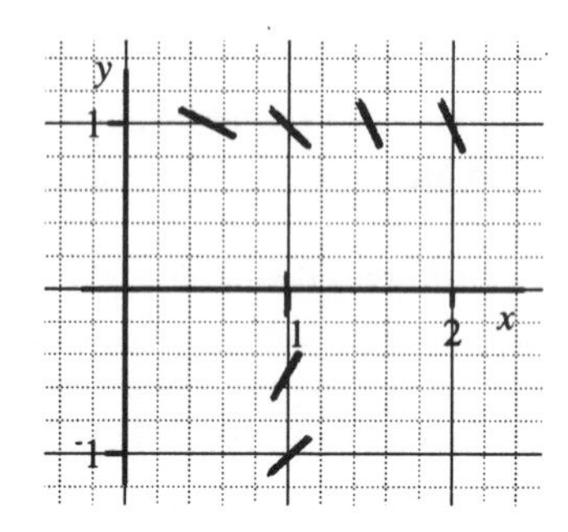

Fig 1.1 Some hand drawn slopes for $\frac{dy}{dx} = -\frac{x}{y}$

In practice we use appropriate computer software to draw direction fields. Figure 1.2 shows the direction field for $-2 < x, y < 2$.

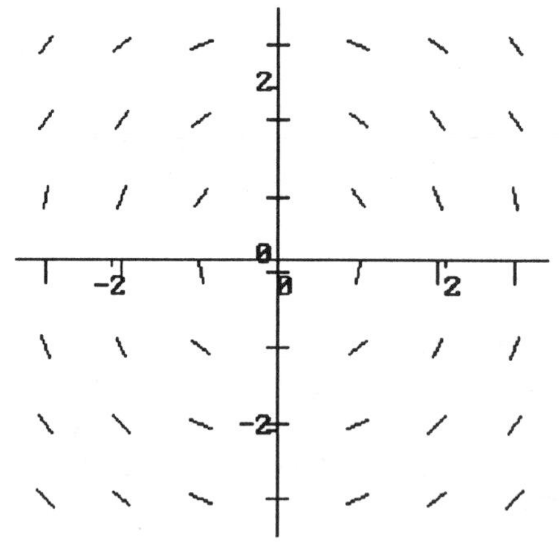

Fig 1.2 Direction field

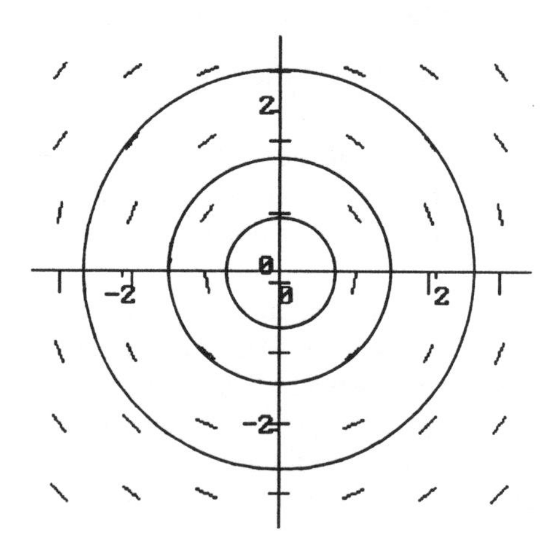

Fig 1.3 Some solution curves

It is immediately obvious from the slopes in Fig 1.2 that the solution curves are circles with centre at the origin. Some of these are shown in Fig 1.3.

The equation of a circle, centre the origin and radius r, is

$$x^2 + y^2 = r^2$$

Differentiating with respect to x gives

$$2x + 2y\frac{dy}{dx} = 0$$

$$\Rightarrow \quad \frac{dy}{dx} = -\frac{x}{y}$$

The equation of a circle satisfies the differential equation and so we deduce that the analytical solution is

$$x^2 + y^2 = r^2$$

For most problems we cannot deduce the analytical solution in this way; furthermore for non-linear differential equations an analytical solution probably does not exist.

Exercises 1.1

1. On graph paper draw the direction field whose slope is 1.5 at every point in the plane. Deduce the equation of the family of curves associated with the field. Check your answer by differentiation.

2. For each of the following differential equations use appropriate software to draw the direction field and sketch (by hand) some of the solution curves:

   (i) $\dfrac{dy}{dx} = -0.5x^2$ (ii) $\dfrac{dy}{dx} = 2y$

   (iii) $\dfrac{dy}{dx} = x + y$ (iv) $\dfrac{dy}{dx} = y - x$

   (v) $\dfrac{dy}{dx} = x^2 + y^2$

3. A parachutist jumps from a hot air balloon. The velocity of the parachutist, v in ms^{-1}, as a function of time t, in seconds, is given from the differential equation

$$\frac{dv}{dt} = 10 - 0.08v^2$$

(i) Draw the direction field.

(ii) From the direction field describe the motion of the parachutist if her initial speed is 5ms^{-1}.

4. The population P of a species of fish in a pond is modelled by the differential equation

$$\frac{dP}{dt} = 0.1P\left(1 - \frac{P}{12\,500}\right)$$

Fig 1.4 shows the direction field of the model.

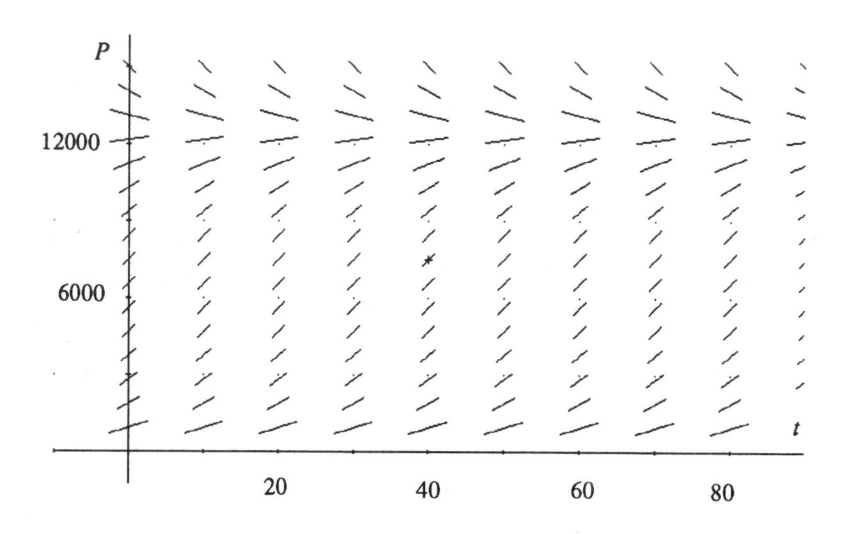

Fig 1.4 The direction field for a population model

(i) On the diagram sketch three solution curves.

(ii) Find the equilibrium population of fish in the pond.

When the population reaches 10 000, fishing from the pond is started. The model now becomes

$$\frac{dP}{dt} = 0.1P\left(1 - \frac{P}{12\,500}\right) - fP$$

where f is a constant. The direction field is now shown in Fig 1.5.

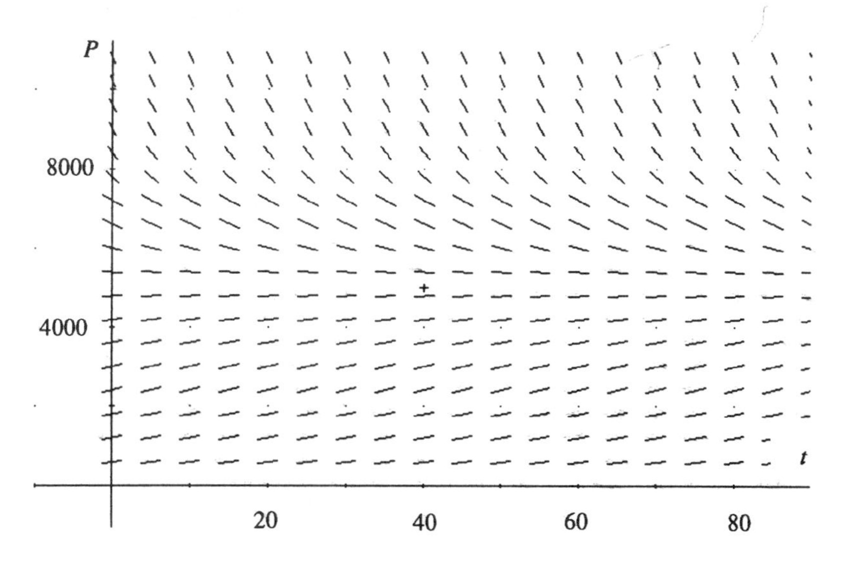

Fig 1.5 Revised model

(iii) On the diagram sketch the solution for P.

(iv) Find the range of values of f for the pond not to be completely emptied of fish.

1.2 First order differential equations: Analytical methods

Many first order differential equations are said to be **separable** if the function $m(x,y)$ can be written as a product (or quotient) of one expression involving x alone and another involving y alone. The differential equation then takes one of the forms

$$\frac{dy}{dx} = f(x)g(y) \quad \text{or} \quad \frac{dy}{dx} = \frac{f(x)}{g(y)} \quad \text{or} \quad \frac{dy}{dx} = \frac{g(y)}{f(x)}$$

The method of solution for such equations is then called **the variables separable method.**

Example 2

Find the general solution of the differential equation

$$\frac{dy}{dx} = -\frac{x}{y}$$

and the particular solution that satisfies the condition $y = 2$ when $x = 1$.

Solution

The differential equation is separable and may be rewritten as

$$y\frac{dy}{dx} = -x$$

Integrating both sides with respect to x leads directly to the general solution

$$\int y \, dy = \int - x \, dx$$

$$\Rightarrow \qquad \tfrac{1}{2}y^2 = -\tfrac{1}{2}x^2 + c$$

where c is a constant of integration.

For the solution to satisfy the condition $y = 2$ when $x = 1$ we require

$$\tfrac{1}{2}(2)^2 = -\tfrac{1}{2}(1)^2 + c \quad \Rightarrow \quad c = \tfrac{5}{2}$$

The particular solution is

$$y^2 = -x^2 + 5$$

The second kind of first order differential equation which we often need to solve is one for which $m(x,y)$ is a linear function of y; so that

$$\frac{dy}{dx} = m(x, y) = f(x)y + g(x)$$

where $f(x)$ and $g(x)$ are functions of x alone. Differential equations of this type are called **linear** and the method of solution involves introducing **an integrating factor**.

Example 3

Find the general solution of the differential equation

$$\frac{dy}{dx} = 2xy + e^{x^2}$$

Solution

This is a linear equation because $m(x,y)$ is linear in y. The integrating factor is a function $p(x)$ defined by

$$p(x) = \exp\left(\int - f(x)\, dx\right) = \exp\left(\int - 2x\, dx\right) = e^{-x^2}$$

On multiplication through by the integrating factor, the differential equation becomes

$$e^{-x^2}\frac{dy}{dx} = 2x\, e^{-x^2} y + 1$$

$$\Rightarrow \quad e^{-x^2}\frac{dy}{dx} - 2x\, e^{-x^2} y = 1$$

The left hand side is the derivative of the product $e^{-x^2} y$, so

$$\frac{d}{dx}\left(e^{-x^2} y\right) = 1$$

Integrating both sides,

$$e^{-x^2} y = x + c$$

$$\Rightarrow \quad y = x\, e^{x^2} + c\, e^{x^2}$$

This is the general solution, where c is a constant.

Note that it is important to write the differential equation in the standard form of equation (1.1) before choosing the appropriate method of solution.

TUTORIAL PROBLEM 1.1 (an important problem for Chapter 2)

Find the particular solution of the differential equation

$$\frac{dx}{dt} = a + bx \qquad b \neq 0$$

which satisfies the initial condition $x = x_0$ when $t = 0$ for the two cases (i) $b > 0$ and (ii) $b < 0$.

In each case describe the long term behaviour of x.

Exercises 1.2

1. Identify which of the following differential equations can be solved by (a) the variables separable method or (b) integrating factor method.

(i) $\frac{dy}{dx} = x - y$ (ii) $y\frac{dy}{dx} = x^2 + 1$

(iii) $\frac{d^2x}{dt^2} - 3\frac{dx}{dt} + 2x = 4e^{3t}$ (iv) $x\frac{dx}{dt} = (t^2 - 1)(x + 4)$

(v) $\frac{dx}{dt} = x - t + 2$ (vi) $\frac{dy}{dt} = 2t + y^2$

2. Find the general solution of the differential equation

$$\frac{dx}{dt} = te^x$$

and the particular solution given that $x = 0$ when $t = 0$.

3. The cooling of an object is modelled by

$$\frac{d\theta}{dt} = -0.05(\theta - 20)$$

where θ is the temperature in °C and t is the time in minutes. If the temperature of the body is initially 80°C find the temperature at any subsequent time t. After how many minutes has the temperature fallen to 50°C?

4. Find the particular solution of each of the following differential equations:

(i) $x\frac{dy}{dx} = y^2$ given $y = 10$ when $x = 1$

(ii) $\frac{dy}{dx} = x^2y^2$ given $y = 2$ when $x = 0$

(iii) $\frac{dy}{dx} = \frac{x^2}{y}$ given $y = 10$ when $x = 0$

5. A small particle moving in a fluid satisfies the differential equation

$$\frac{dv}{dt} = -0.2(v + v^2)$$

Find the particular solution for $v(t)$ given that the initial speed of the particle is 40 ms^{-1}.

6. Find the general solution of each of the following differential equations:

(i) $\frac{dy}{dx} = 3x - 4y$ (ii) $x\frac{dy}{dx} = y - x$

(iii) $\frac{dy}{dx} = y + e^x$ (iv) $x\frac{dy}{dx} + 3y = 5x^2$

(v) $\frac{dx}{dt} = 2x + 5t$

1.3 Second order differential equations

In Chapter 3 you will only need to solve second order homogeneous differential equations with constant coefficients.

Example 4

Find the general solution of the differential equation

$$\frac{d^2y}{dx^2} - 2\frac{dy}{dx} + 5y = 0$$

Solution

The differential equation is second order because the order of the highest derivative is 2. A solution of this equation is of the form $e^{\lambda x}$ where λ is a constant to be determined.

If we let $y = e^{\lambda x}$ then $\dfrac{dy}{dx} = \lambda e^{\lambda x}$ and $\dfrac{d^2 y}{dx^2} = \lambda^2 e^{\lambda x}$. Substituting for these into the differential equation gives

$$\lambda^2 e^{\lambda x} - 2\lambda e^{\lambda x} + 5e^{\lambda x} = 0$$

Dividing by $e^{\lambda x}$ gives the **auxiliary equation**

$$\lambda^2 - 2\lambda + 5 \Rightarrow \lambda = 1 \pm 2i$$

The general solution of the differential equation is then

$$y = e^x(A\cos 2x + B\sin 2x)$$

For the general second order equation

$$a\frac{d^2y}{dx^2} + b\frac{dy}{dx} + cy = 0 \tag{1.2}$$

the auxiliary equation is

$$a\lambda^2 + b\lambda + c = 0 \tag{1.3}$$

If equation (1.3) has two distinct real roots λ_1 and λ_2 then the general solution of equation (1.2) is

$$y = Ae^{\lambda_1 x} + Be^{\lambda_2 x}$$

If equation (1.3) has one repeated root λ then the general solution of equation (1.2) is

$$y = (A + Bx)e^{\lambda x}$$

If equation (1.3) has complex conjugate solutions, $\lambda_1 = \alpha + i\beta$ and $\lambda_2 = \alpha - i\beta$ then the general solution of equation (1.2) is

$$y = e^{\alpha x}(A\cos \beta x + B\sin \beta x)$$

In each case A and B are arbitrary constants.

Exercises 1.3

Find the general solution for each of the following differential equations:

(i) $5\frac{d^2y}{dx^2} - 6\frac{dy}{dx} + y = 0$

(ii) $4\frac{d^2x}{dt^2} + x = 0$

(iii) $\frac{d^2y}{dx^2} - 2\frac{dy}{dx} + y = 0$

(iv) $\frac{d^2x}{dt^2} + \frac{dx}{dt} + x = 0$

(v) $\frac{d^2y}{dx^2} + 2\frac{dy}{dx} - 3y = 0$

1.4 Taylor polynomials

Taylor polynomials provide an approximation of a function, in a neighbourhood of some given point, in terms of polynomials whose coefficients are given in terms of the derivatives of the function.

In the study of many non-linear systems, the approximating linear Taylor polynomial will provide an important step in the analysis. We shall see that linear systems are fairly easy to describe qualitatively and quantitatively. The same is not true of non-linear systems. So the link between a non-linear function and its Taylor polynomials will be our lifeline. We shall restrict the review to second order Taylor polynomials.

The form for the second Taylor polynomial for a function of one variable $f(x)$ about $x = a$ is

$$f(a) + (x-a)\frac{df}{dx}(a) + \frac{(x-a)^2}{2}\frac{d^2f}{dx^2}(a)$$

For a function of two variables $f(x,y)$ about (a,b) the second Taylor polynomial is

$$f(a,b) + (x-a)\frac{\partial f}{\partial x}(a,b) + (y-b)\frac{\partial f}{\partial y}(a,b) + \frac{(x-a)^2}{2}\frac{\partial^2 f}{\partial x^2}(a,b)$$
$$+ (x-a)(y-b)\frac{\partial^2 f}{\partial x \partial y}(a,b) + \frac{(y-b)^2}{2}\frac{\partial^2 f}{\partial y^2}(a,b)$$

Higher order Taylor polynomials are sometimes needed if the first and second order derivatives are zero at the point of interest.

Example 5

Find the second Taylor polynomial for the function $f(x) = e^{-x^2}$ about the point $x = 1$.

Solution

For the second Taylor polynomial we need the first and second derivative of f evaluated at $x = 1$.

$$f(x) = e^{-x^2} \Rightarrow \frac{df}{dx} = -2xe^{-x^2} \qquad \Rightarrow \frac{df}{dx}(1) = -2e^{-1}$$

$$\frac{d^2f}{dx^2} = -2e^{-x^2} + 4x^2e^{-x^2} \Rightarrow \frac{d^2f}{dx^2}(1) = 2e^{-1}$$

The second Taylor polynomial for $f(x) = e^{-x^2}$ about $x = 1$ is

$$e^{-1} - 2e^{-1}(x-1) + e^{-1}(x-1)^2$$

Example 6

Find the second Taylor polynomial for the function $f(x,y) = \cos(4x + 3y)$ about the point $\left(\frac{\pi}{8}, \frac{\pi}{6}\right)$.

Solution

For the function itself, $4x+3y = 4\left(\frac{\pi}{8}\right) + 3\left(\frac{\pi}{6}\right) = \pi$, so $f\left(\frac{\pi}{8}, \frac{\pi}{6}\right) = \cos\pi = -1$. For the second Taylor polynomial we need the first and second partial derivatives of f evaluated at $\left(\frac{\pi}{8}, \frac{\pi}{6}\right)$.

$$\frac{\partial f}{\partial x} = -4\sin(4x+3y); \qquad \text{at } \left(\frac{\pi}{8}, \frac{\pi}{6}\right) \quad \frac{\partial f}{\partial x} = -4\sin\pi = 0$$

$$\frac{\partial f}{\partial y} = -3\sin(4x+3y); \qquad \text{at } \left(\frac{\pi}{8}, \frac{\pi}{6}\right) \quad \frac{\partial f}{\partial y} = -3\sin\pi = 0$$

$$\frac{\partial^2 f}{\partial x^2} = -16\cos(4x+3y); \qquad \text{at } \left(\frac{\pi}{8}, \frac{\pi}{6}\right) \quad \frac{\partial^2 f}{\partial x^2} = -16\cos\pi = 16$$

$$\frac{\partial^2 f}{\partial x \partial y} = -12\cos(4x+3y); \quad \text{at} \quad \left(\frac{\pi}{8},\frac{\pi}{6}\right) \quad \frac{\partial^2 f}{\partial x \partial y} = -12\cos\pi = 12$$

$$\frac{\partial^2 f}{\partial y^2} = -9\cos(4x+3y); \quad \text{at} \quad \left(\frac{\pi}{8},\frac{\pi}{6}\right) \quad \frac{\partial^2 f}{\partial y^2} = -9\cos\pi = 9$$

The second Taylor polynomial for $f(x,y) = \cos(4x + 3y)$ about the point $\left(\frac{\pi}{8},\frac{\pi}{6}\right)$ is

$$-1+\left(x-\frac{\pi}{8}\right)(0)+\left(y-\frac{\pi}{6}\right)(0)+\left(x-\frac{\pi}{8}\right)^2\frac{(16)}{2}+\left(x-\frac{\pi}{8}\right)\left(y-\frac{\pi}{6}\right)(12)+\left(y-\frac{\pi}{6}\right)^2\frac{(9)}{2}$$

$$= -1+8\left(x-\frac{\pi}{8}\right)^2 + 12\left(x-\frac{\pi}{8}\right)\left(y-\frac{\pi}{6}\right)+\frac{9}{2}\left(y-\frac{\pi}{6}\right)^2$$

Exercises 1.4

Find the second Taylor polynomial for each of the following functions about the given point:

(i) $f(x) = \sin x$ about $x = 0$

(ii) $f(x) = x^3$ about $x = 1$

(iii) $f(x) = \frac{1}{x}$ about $x = -2$

(iv) $f(x) = \sqrt{x+1}$ about $x = 0$

(v) $f(x,y) = 1+x+\frac{y^2}{x}$ about $(1,2)$

(vi) $f(x,y) = 2+xy^2-\frac{x}{y}$ about $(1,-1)$

(vii) $f(x,y) = \sin(xy)$ about $\left(0,\frac{\pi}{4}\right)$

(viii) $f(x,y) = \ln(x + y - 1)$ about $(0,1)$

1.5 Algebra of matrices

In Chapter 3 the analysis of second order systems uses properties of 2×2 matrices. In this section we review some of their basic properties.

Example 7

For the matrix $\mathbf{A} = \begin{bmatrix} 3 & 1 \\ 2 & 4 \end{bmatrix}$

(i) evaluate $\mathrm{tr}(\mathbf{A})$, $\det(\mathbf{A})$ and $\mathbf{A}^{-1}$,

(ii) find the eigenvalues and eigenvectors of $\mathbf{A}$.

Solution

(i) The **trace** of a matrix, denoted by $\mathrm{tr}(\mathbf{A})$, is the sum of the elements on the diagonal. In this case

$$\mathrm{tr}(\mathbf{A}) = 3 + 4 = 7$$

$$\det(\mathbf{A}) = \begin{vmatrix} 3 & 1 \\ 2 & 4 \end{vmatrix} = 3 \times 4 - 1 \times 2 = 10$$

The **inverse** of a matrix $\mathbf{A}$, denoted by $\mathbf{A}^{-1}$ exists if $\det(\mathbf{A}) \neq 0$ and is such that $\mathbf{A}\mathbf{A}^{-1} = \mathbf{A}^{-1}\mathbf{A} = \mathbf{I}$. For a 2×2 matrix the inverse is easily found using the following:

$$\text{If } \mathbf{A} = \begin{bmatrix} a & b \\ c & d \end{bmatrix} \text{ then } \mathbf{A}^{-1} = \frac{1}{\det(\mathbf{A})} \begin{vmatrix} d & -b \\ -c & a \end{vmatrix}$$

In this case,

$$\mathbf{A}^{-1} = \frac{1}{10} \begin{vmatrix} 4 & -1 \\ -2 & 3 \end{vmatrix}$$

If $\mathbf{A}^{-1}$ does not exist then $\mathbf{A}$ is said to be **singular**, otherwise it is **non-singular**.

(ii) An **eigenvalue** of a square matrix $\mathbf{A}$ is a number λ for which the equation

$$\mathbf{A}\mathbf{x} = \lambda\mathbf{x}$$

has non-zero solutions for the vector $\mathbf{x}$. The corresponding vector $\mathbf{x}$ for an eigenvalue is called an **eigenvector**. In this case

$$\begin{bmatrix} 3 & 1 \\ 2 & 4 \end{bmatrix} \mathbf{x} = \lambda \mathbf{x}$$

can be rewritten as

$$\begin{bmatrix} 3-\lambda & 1 \\ 2 & 4-\lambda \end{bmatrix} \mathbf{x} = \mathbf{0}$$

This has non-zero solutions for **x** provided the determinant

$$\begin{vmatrix} 3-\lambda & 1 \\ 2 & 4-\lambda \end{vmatrix} = 0$$

$\Rightarrow \quad (3-\lambda)(4-\lambda) - 2 = 0$

$\Rightarrow \quad \lambda^2 - 7\lambda + 10 = 0$

$\Rightarrow \quad (\lambda - 5)(\lambda - 2) = 0$

$\Rightarrow \quad \lambda = 5 \quad \text{or} \quad \lambda = 2$

The eigenvalues are $\lambda = 5$ and $\lambda = 2$.

For $\lambda = 5$, the equations for the eigenvectors are

$$\begin{bmatrix} 3-5 & 1 \\ 2 & 4-5 \end{bmatrix} \mathbf{x} = \mathbf{0}$$

If we denote **x** by $[u \;\; v]^{-1}$ then expanding the left hand side gives

$$\begin{aligned} -2u + v &= 0 \\ 2u - v &= 0 \end{aligned}$$

with solutions $\alpha[1 \;\; 2]^T$ for any constant α.

For $\lambda = 2$ the equations for the eigenvectors are

$$\begin{bmatrix} 3-2 & 1 \\ 2 & 4-2 \end{bmatrix} \mathbf{x} = \mathbf{0}$$

$\Rightarrow \quad u + v = 0$

with solutions $\beta[1 \;\; -1]^{\mathrm{T}}$ for any constant β.

Matrices have all sorts of uses and applications in the geometry of space. For example the matrix

$$\mathbf{R} = \begin{bmatrix} \cos\theta & \sin\theta \\ -\sin\theta & \cos\theta \end{bmatrix}$$

represents a rotation through an angle θ of the coordinate axes in an anti-clockwise sense about the origin. The axes of the new coordinate system (x',y') are still orthogonal. The geometry of $\mathbf{R}$ is shown in Fig 1.6.

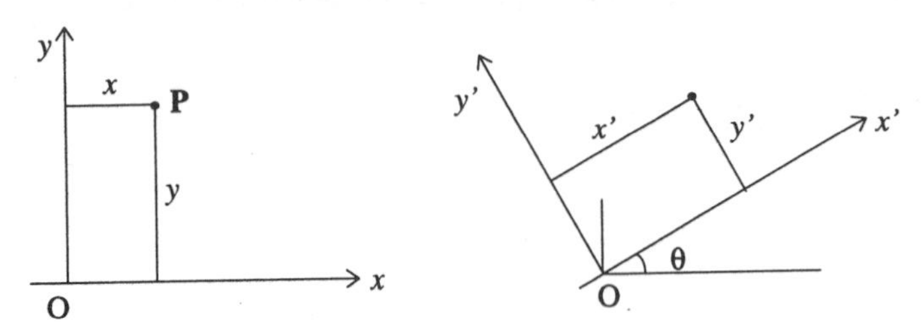

Fig 1.6 The matrix **R** is a rotation through an angle θ

For a general non-singular 2×2 matrix $\mathbf{T}$ the transformation of coordinates given by $\mathbf{x}' = \mathbf{T}\mathbf{x}$ produces an oblique set of coordinate axes in the $\mathbf{x}'$ plane. The geometry of $\mathbf{T}$ is shown in Fig 1.7.

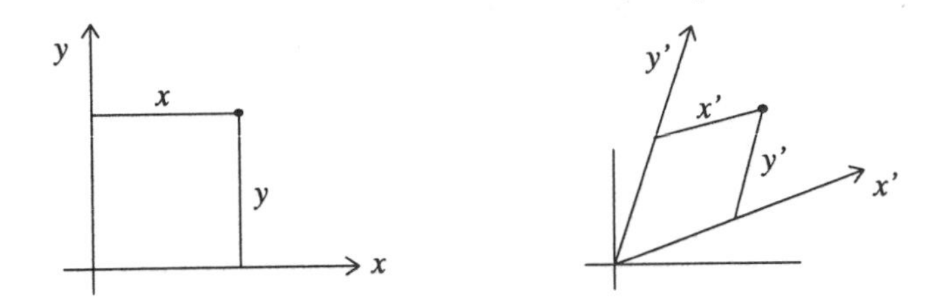

Fig 1.7 A general non-singular 2 × 2 matrix **T** forms oblique coordinate axes

Exercises 1.5

1. For each of the following 2×2 matrices

   (i) evaluate $\text{tr}(\mathbf{A})$, $\det(\mathbf{A})$ and $\mathbf{A}^{-1}$,
   (ii) find the eigenvalues and eigenvectors of $\mathbf{A}$.

   (a) $\mathbf{A} = \begin{bmatrix} -2 & -2 \\ -1 & -3 \end{bmatrix}$ (b) $\mathbf{A} = \begin{bmatrix} \sqrt{3} & 1 \\ -1 & \sqrt{3} \end{bmatrix}$ (c) $\mathbf{A} = \begin{bmatrix} 3 & 9 \\ 0 & 2 \end{bmatrix}$

   (d) $\mathbf{A} = \begin{bmatrix} 0 & 1 \\ 1 & 0 \end{bmatrix}$ (e) $\mathbf{A} = \begin{bmatrix} 1 & -1 \\ -1 & 1 \end{bmatrix}$ (f) $\mathbf{A} = \begin{bmatrix} 1 & 5 \\ -2 & 3 \end{bmatrix}$

2. Consider the transformation from Cartesian coordinates $\mathbf{x}$ to oblique coordinates $\mathbf{x}'$ given by $\mathbf{x}' = \mathbf{A}\mathbf{x}$. For each of the following matrices $\mathbf{A}$ find the eigenvectors and show that the direction of the eigenvectors is along the oblique coordinate axes.

(i) $\mathbf{A} = \begin{bmatrix} 5 & 2 \\ 3 & 4 \end{bmatrix}$ (ii) $\mathbf{A} = \begin{bmatrix} 0 & 1 \\ 4 & 0 \end{bmatrix}$

3. Show that the following two matrices have one eigenvalue λ. Show that $\mathbf{A}$ has two linearly independent eigenvectors while for $\mathbf{B}$ every eigenvector is a scalar multiple of $[0 \;\; 1]^T$.

$$\mathbf{A} = \begin{bmatrix} \lambda & 0 \\ 0 & \lambda \end{bmatrix} \qquad \mathbf{B} = \begin{bmatrix} \lambda & 0 \\ 1 & \lambda \end{bmatrix}$$

1.6 Iteration and cobweb diagrams

In Chapter 4 we investigate non-linear systems described by discrete models and, in particular, difference equations (or recurrence relations). The equation

$$x_{n+1} = 0.2x_n - 2$$

is an example of a **difference equation** of first order. It is an equation that can be used repeatedly to generate a sequence of numbers once the initial term is known. The process of continually repeating an equation of this type is called **iteration**.

Example 8

Investigate the sequence obtained by iterating the following two difference equations starting with $x_0 = 0$:

(i) $x_{n+1} = 0.2x_n + 2$
(ii) $x_{n+1} = 2x_n + 2$

Solution

The following table shows the two sequences:

	n	0	1	2	3	4	5	6	7	8	9	10
(i)	x_n	0	2	2.4	2.48	2.496	2.4992	2.499 84	2.499 96	2.5	2.5	2.5
(ii)	x_n	0	2	6	14	30	62	126	254	510	1022	2046

We see that the sequence for the difference equation $x_{n+1} = 0.2x_n + 2$ converges to the value 2.5, whereas $x_{n+1} = 2x_n + 2$ diverges.

In general the linear recurrence relation $x_{n+1} = ax_n + p$ converges to a limit if $|a| < 1$ and diverges if $|a| \geq 1$.

A geometrical representation of the sequence generated by a difference equation that will help us understand the dynamics of a discrete system is the **cobweb** or **staircase** diagram. Consider the sequence generated by the equation

$$x_{n+1} = 0.8x_n + 2$$

starting with $x_0 = 0$. Figure 1.8 shows a graph of the straight lines $y = 0.8x + 2$ and $y = x$ and the staircase diagram showing the sequence generated by iteration.

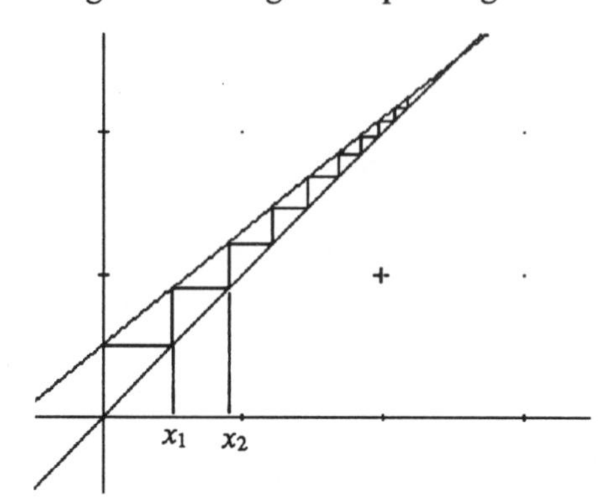

Fig 1.8 Staircase diagram for $x_{n+1} = 0.8x_n + 2$

We begin the staircase on the diagonal line $y = x$ at the point (x_0,x_0) (in this case we start at the origin (0,0)). We move vertically to the graph of $y = 0.8x + 2$ and then horizontally to the diagonal line. We reach the diagonal line at the next iterate x_1. Now we repeat the process, e.g. vertically to the line $y = 0.8x + 2$ then horizontally to the diagonal line, to obtain the iterate x_2 and so on.

Note that to obtain a staircase or cobweb diagram that is meaningful for the interpretation of a difference equation it is important to draw the graphs accurately, i.e. not the usual sketch graphs!

In the next example we illustrate a cobweb diagram.

Example 9

Show graphically the iterative process for the difference equation

$$x_{n+1} = \cos x_n$$

with initial value $x_0 = 0$.

Solution

Figure 1.9 shows graphs of $y = x$ and $y = \cos x$ and the sequence generated by the geometrical iterative process of 'moving vertically to $y = \cos x$ and horizontally to $y = x$'.

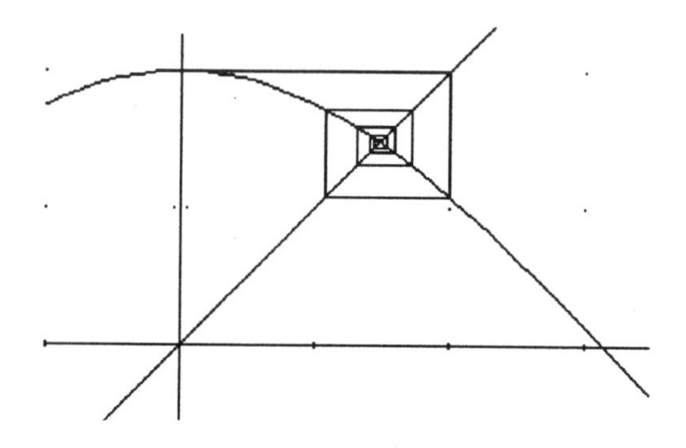

Fig 1.9 Cobweb diagram for $x_{n+1} = \cos x_n$

Exercises 1.6

1. Investigate the sequence obtained by iterating each of the following difference equations for the given initial value. Decide whether or not the sequence of iterations converges or diverges.

   (i) $x_{n+1} = 0.5x_n + 1.5$ $\quad x_0 = 1$

   (ii) $x_{n+1} = x_n^{1.1} + 0.1$ $\quad x_0 = 1$

   (iii) $x_{n+1} = 1.5x_n - 2$ $\quad x_0 = 3$

   (iv) $x_{n+1} = 1 + \sin x_n$ $\quad x_0 = 0$

   (v) $x_{n+1} = 2 + 1.5\sqrt{x_n}$ $\quad x_0 = 2$

2. Show that the general solution of the equation

$$x_{n+1} = ax_n$$

is $x_n = Aa^n$ where A is the starting value of x_n.

3. Show graphically the iterative process for each of the difference equations in problem 1. Confirm your decision about the convergence or divergence of the sequence of iterations.

2 •First Order Continuous Autonomous Systems

2.1 Some terminology

Inevitably at the start of a topic in mathematics we begin with some of the special terminology associated with the topic. This book is about the motion of systems. By **system** we mean an object or set of objects under investigation. For example, the system could be a biological system such as the trout in a fish farm. In modelling such a system we would be interested in how the number of trout varies 'naturally' from season to season and in the effects of fishing.

The system could represent an environmental problem such as modelling the effect on communities living near a motorway of the lead in the atmosphere. The lead reaches the bones in a human body through the blood and tissue. Too much lead in a human can produce serious physical disabilities.

The system could represent a physical situation such as the motion of a stone thrown through a window, the motion of the planets round the sun or the temperature of a cooling body.

The first step in describing all of these situations mathematically is to find a mathematical model (or models) consisting of a set of assumptions and simplifications and one or two variables which represent the features we are interested in. Take for example the modelling of a cooling cup of coffee. The temperature of the coffee, T, at time t is given by Newton's law of cooling

$$\frac{\mathrm{d}T}{\mathrm{d}t} = -k(T - T_0)$$

where k is constant and T_0 is the temperature of the surrounding air. This model is based on experiment and assumes that the heat loss varies linearly with the excess temperature above the surroundings.

The systems that we model are often called **dynamical systems** to imply **change**. (Think of dynamics, which is a subject within the field of mechanics, meaning the motion of objects; this is different from statics, when the system under investigation is at rest.)

In this chapter the mathematical model describing each system will be a first order differential equation of the form

$$\frac{\mathrm{d}x}{\mathrm{d}t} = v(x, t)$$

where x is the dependent variable and t is the independent variable, usually representing time.

The function $v(x,t)$ is called the **velocity field** of the system and its value for particular values of x and t is the **phase velocity**. This should not be confused with the notion of velocity in mechanics. For example in modelling the motion of a parachutist dropping vertically, the equation of motion is

$$\frac{dv}{dt} = -g - kv^2$$

where g is the acceleration due to gravity and kv^2 is the air resistance per unit mass. In this model v is the velocity of the parachutist and t is the time. So the 'velocity field' for this dynamical system is $-g - kv^2$ which is the acceleration of the parachutist. In modelling problems in mechanics we must be very careful with the interpretation of the 'velocity field' and 'phase velocity'. For the coffee cooling problem the velocity field is

$$v(T,t) = -k(T - T_0)$$

and clearly in this case there can be no confusion.

If the velocity field does not depend explicitly on time, so that $v(x,t)$ is a function of x only, then the system is called an **autonomous system**. Otherwise the system is non-autonomous.

TUTORIAL PROBLEM 2.1

Sketch the direction field for a general autonomous system

$$\frac{dx}{dt} = v(x)$$

showing the symmetry that you would expect. How can you tell from inspecting a direction field whether or not a velocity function is autonomous?

TUTORIAL PROBLEM 2.2

Show that the solution of an autonomous system

$$\frac{dx}{dt} = v(x)$$

with initial condition $x = x_0$ when $t = t_0$ depends on the difference $t - t_0$.

Since the velocity field is a function of x only, the direction field is independent of t so that at all points with the same x value the tangent lines will all be parallel (see Fig 2.1).

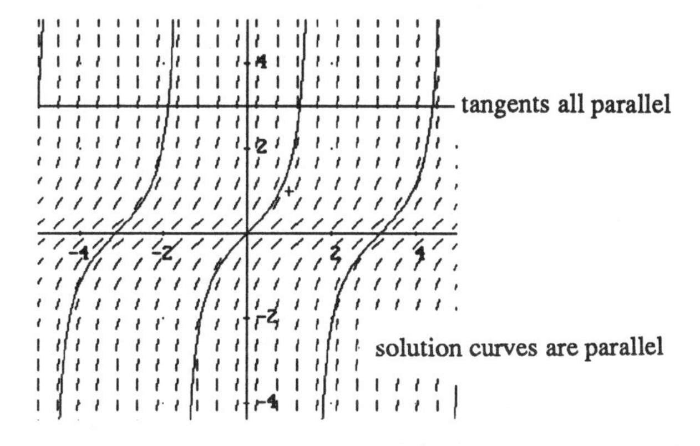

Fig 2.1 The direction field for an autonomous system

Hence the solution curves will all be parallel, i.e. they are unchanged by a translation parallel to the t-axis. To identify an autonomous system from the direction field we look for parallel tangent lines for each x value. For an autonomous system we can formally integrate the differential equation because it is of the variables separable type. Suppose that $x = x_0$ when $t = t_0$; then

$$\int_{x_0}^{x} \frac{1}{v(x)} dx = \int_{t_0}^{t} 1 dt = (t - t_0)$$

The left hand side of this equation is a function of x only. Formally assuming that we can integrate $\frac{1}{v(x)}$ and then find the inverse function, we can see that the solution for x will be a function of the difference $t - t_0$. For example, suppose that $v(x) = \frac{1}{x}$; then

$$\int_{x_0}^{x} \frac{1}{v(x)} dx = \int_{x_0}^{x} x\, dx = \frac{x^2}{2} - \frac{x_0^2}{2} = t - t_0$$

Rearranging and solving for x,

$$x = \sqrt{x_0^2 + 2(t - t_0)}$$

Figure 2.2 shows graphs of this solution for $x_0 = 1$, $t_0 = 0$ and $x_0 = 1$, $t_0 = 2$.

What this solution implies is that for an autonomous system, two solution curves that pass through the same point x_0 (= 1 in Fig 2.2) a time T apart ($T = 2$ in Fig 2.2) will always be separated by the same time T.

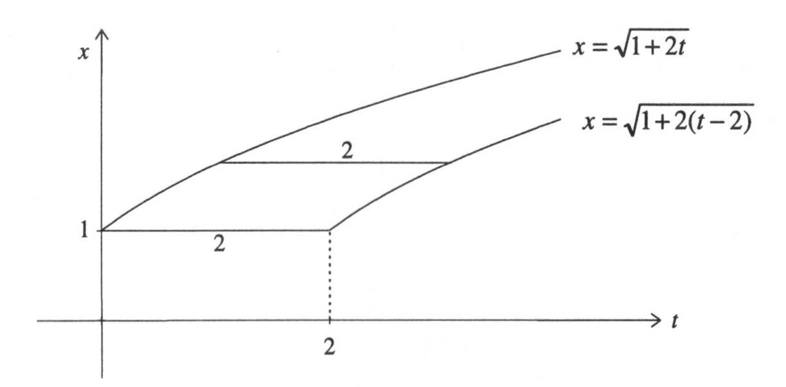

Fig 2.2 Translating a solution curve

The **state** of the system under investigation is specified by giving the dependent variable x a specific value. For example, the state of the cooling coffee system is determined by the temperature of the coffee at a given time.

To describe the state of an autonomous dynamical system over an interval of time we need to solve the equation $\frac{dx}{dt} = v(x)$ to find a solution $x(t)$. The state of the system at some given time t_0 is given by $x(t_0)$ which can be represented by a point on the x-axis, called the **phase point**. As time increases the state of the dynamical system changes and the point representing the system moves along the x-axis with velocity $v(x)$. Thus the changes of the system are represented by the motion of the phase point along the x-axis which is often called the **phase line**.

Example 1

A can of cold drink is taken from the fridge at a temperature of 5°C and left to warm up in a kitchen where the air temperature is 20°C. Draw the phase line for this system assuming that the model is 'Newton's law of cooling'.

Solution

Let the temperature of the cold drink at time t seconds after being removed from the fridge be T°C. Then Newton's law of cooling gives

$$\frac{dT}{dt} = -k(T-20)$$

Since in this case $T < 20$, $\frac{dT}{dt} > 0$ so that the temperature increases as expected.

For this system the dependent variable is T and the velocity field is

$$v(T) = -k(T-20)$$

This is an autonomous system. Suppose we choose two values of T, say $T = 5°C$ and $T = 10°C$. Then $v(5) = -k(5 - 20) = 15k$ and $v(10) = -k(10 - 20) = 10k$. The phase portraits are shown in Fig 2.3. The velocity is represented by an arrow of length proportional to the magnitude of the velocity.

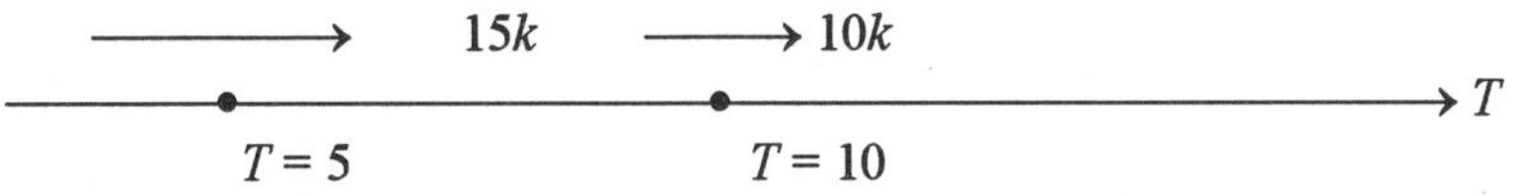

$15k$ $10k$ T

$T = 5$ $T = 10$

Fig 2.3 Phase portraits for $\dfrac{dT}{dt} = -k(T - 20)$

Note that if in the above example the drink was cooling from a temperature above 20°C then $v(T) < 0$. For example, if $T = 30°C$, $v(30) = -10k$ and the phase portrait is shown in Fig 2.4 with the arrow pointing to the left.

$10k$ T

Fig 2.4 Phase portrait for $T > 20°C$

The direction of the arrow in the phase portrait shows the direction of motion of the dynamical system. From this direction we can deduce the future changes of the system. If the arrow points to the right then the dependent variable will increase in size, if the arrow points to the left the dependent variable will decrease in size.

Sometimes there are values of the dependent variable for which the velocity field is zero. For example, if $T = 20$ in Example 1 then $v(20) = 0$. This is represented on the phase line by a point and no arrow. Such a point is called a **fixed point** (see Fig 2.5).

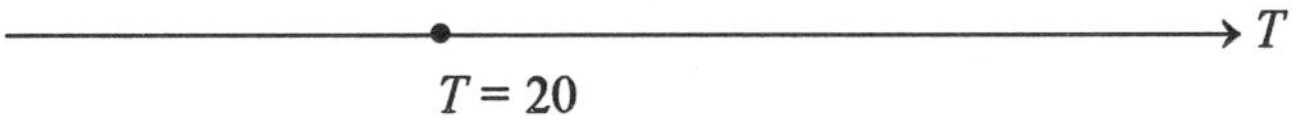

T

$T = 20$

Fig 2.5 The phase portrait for a fixed point

Exercises 2.1

1. Which of the following velocity functions are autonomous?

(i) $v(x,t) = e^x$ (ii) $v(x,t) = \sin 4t$ (iii) $v(x,t) = e^{x-t}$

(iv) $v(x,t) = x^3$ (v) $v(x,t) = \begin{cases} 0 & t < 0 \\ x^2 & t \geq 0 \end{cases}$

2. The following diagrams show the direction field and some solution curves for four dynamical systems. Which graphs could represent autonomous systems?

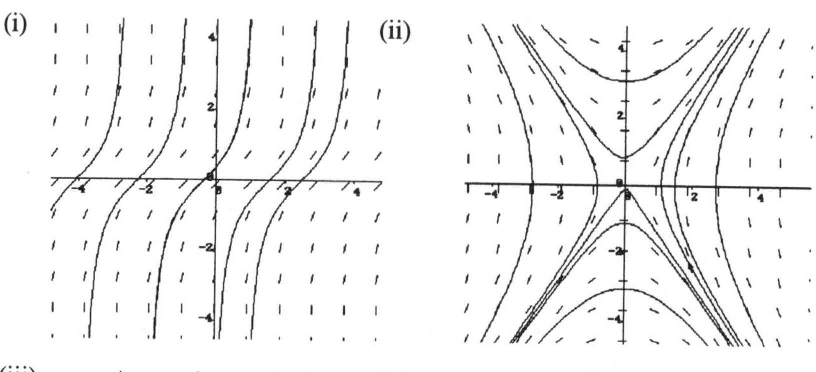

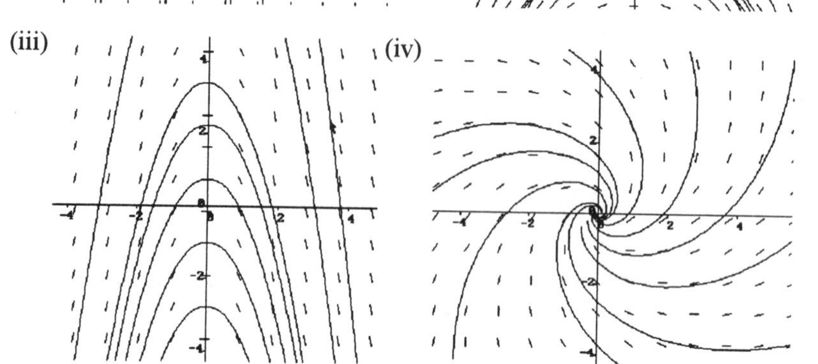

3. Find the solution of the autonomous equation

$$\frac{dx}{dt} = x$$

with initial conditions $x = x_0$ when $t = t_0$. Show that the solution satisfies the property of Tutorial Problem 2.2.

4. Find the solution of the non-autonomous equation

$$\frac{dx}{dt} = xt$$

with initial conditions $x = x_0$ when $t = t_0$. Show that the solution depends on both t and t_0 and not just on the difference $t - t_0$.

5. Sketch the direction field for a general non-autonomous system

$$\frac{dx}{dt} = v(t)$$

showing the symmetry that you would expect. What familiar property of integration is associated with this symmetry?

6. The motion of a stone dropped vertically down a well is modelled by the equation

$$\frac{du}{dt} = 10 - 0.2u^2$$

where u is the speed of the stone at time t. Draw the phase portraits for different values of u. Find the fixed point for this dynamical system.

7. For a chemical reaction of two substances, if $N(t)$ is the concentration of one of the substances at time t then the reaction is modelled by the equation

$$\frac{dN}{dt} = k(a - N)(b - N)$$

where a, b and k are positive constants. Draw phase portraits for different values of N in the intervals (i) $N < a < b$, (ii) $a < N < b$ and (iii) $a < b < N$. Find the fixed points for this dynamical system.

2.2 Classification of the fixed points of autonomous systems

In the previous section we introduced the idea of a **fixed point** defined as a value of $x = x_f$ for which $v(x_f) = 0$. At a fixed point, $\frac{dx}{dt} = 0$ and a system initially at a zero of $v(x)$ remains there for all time. Sometimes a fixed point is called an **equilibrium point** so that a system at a fixed point is said to be in equilibrium.

At a fixed point a system does not change, so that when modelling real systems fixed points are an important part of the description. Consider the warming of the cold drink in Example 1. The model is Newton's law of cooling,

$$\frac{dT}{dt} = -k(T - 20)$$

The fixed point of the system is $T = 20$°C. If the drink is at this temperature then it neither warms up nor cools down; it is in equilibrium. Either side of this equilibrium temperature the drink will cool if $T > 20$°C and will warm up if $T < 20$°C. Whatever we do to the drink it will attempt to reach the air temperature of 20°C. We could deduce that the system is **stable** in that following a small change in temperature from 20°C the drink will return to equilibrium.

This description is typical of the three features of a system that we are interested in:

- Where are the fixed points or equilibrium points?
- What happens if the system is displaced from equilibrium?
- What is the long term behaviour of the system?

The phase portrait of a system is an important part of answering these questions. We have seen that for an autonomous system the solution curves are all parallel so the direction field contains the same information over and over again and provides an unnecessarily complicated view in this case.

To represent the velocity field of an autonomous system, at a selected number of points on the x-axis, we draw arrows with the following properties:

1. the length of the arrow is proportional to $|v(x)|$;
2. the direction of the arrow is to the right (x increasing) if $v(x) > 0$ and to the left (x decreasing) if $v(x) < 0$.

In a sense this is a one-dimensional direction field. It is called the **phase portrait** or **phase diagram**.

Example 2

Draw the phase portrait of the dynamical system

$$\frac{dx}{dt} = x(x^2 - 1)$$

by choosing the points $x = -1.5, -1, -0.5, 0.5, 1, 1.5$. Use the phase portrait to describe the motion of the system.

Solution

In this example the velocity field $v(x) = x(x^2 - 1)$ and the following table shows the values of v at the given phase points.

x	-1.5	-1	-0.5	0	0.5	1	1.5
$v(x)$	-1.875	0	0.375	0	-0.375	0	1.875

We see that there are fixed points at the phase points $x = -1$, $x = 0$ and $x = 1$. Figure 2.6 shows the phase portrait.

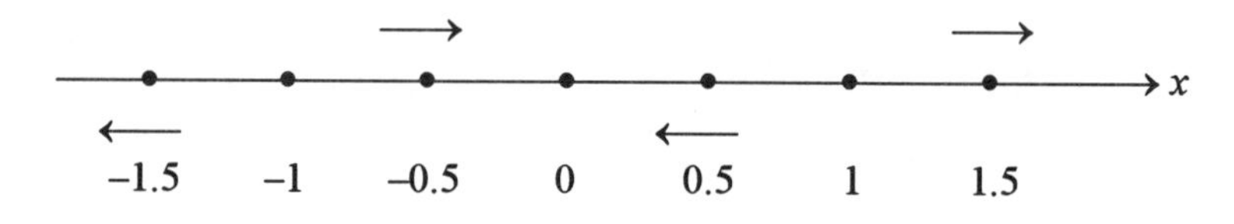

Fig 2.6 Phase portrait for $v(x) = x(x^2 - 1)$

The phase portrait can be used to describe the motion of the system qualitatively. For instance, a point initially at $x = -1.5$ is moving away from the fixed point at $x = -1$. A point initially at $x = -0.5$ will move away from the fixed point at $x = -1$ and towards the fixed point at the origin. Similarly a point initially at $x = 0.5$ will move towards the origin and a point initially at $x = 1.5$ will move away from the fixed point at $x = 1$. If the initial point is at one of the fixed points then $\frac{dx}{dt} = v(x) = 0$ and the point stays where it is. A phase point initially on one side of a fixed point, say $x = 0$, cannot pass through to the other side of the fixed point because at the fixed point $v(x) = 0$.

What happens to the left of $x = -1.5$ and to the right of $x = 1.5$? To answer this question a graph of $v(x)$ against x is often helpful (see Fig 2.7).

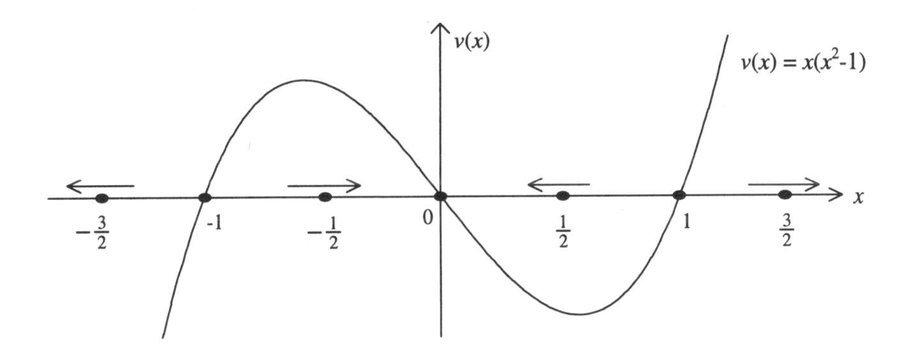

Fig 2.7 Phase portrait and a graph of $v(x)$

From the graph of $v(x)$ (Fig 2.7) it is easy to deduce the sign and size of the velocity field and then to deduce the qualitative behaviour of the system.

To the left of the fixed point at $x = -1$, $v(x) < 0$ and $|v(x)|$ is increasing indefinitely. We can deduce that the point moves away from the fixed point with increasing speed. However, just to the right of the fixed point at $x = -1$ the point gains speed until it reaches a maximum at $x = -0.5$ and then moves slower and slower as it approaches the origin. Similarly we can describe the behaviour for initial points to the right of the origin.

Since a phase point cannot move through a fixed point, the x-axis divides itself naturally into intervals called **invariant sets**. Given an initial position of the system then the phase point is captured within an invariant set. In Example 2 the invariant sets are $(-\infty,-1)$, $(-1,0)$, $(0,1)$ and $(1,\infty)$. Note that the invariant sets are open intervals. A point which is initially inside the interval $(-1,0)$ will remain there.

TUTORIAL PROBLEM 2.3

Draw the phase portrait of the dynamical system

$$\frac{dx}{dt} = x(2-x)$$

by choosing the points $x = -1, 0, 1, 2$ and 3. Describe the motion of the system. Write down the invariant sets for this system.

An interesting question is how much information is lost from the qualitative description obtained from the phase portrait as opposed to the full analytical solution? Consider the simple linear system

$$\frac{dx}{dt} = x$$

This differential equation can be solved using the variables separable method

$$\int \frac{1}{x} dx = dt$$

$$\ln |x| = t + c$$

$$x = Ae^t$$

(where we have written $A = e^c$).

Consider various initial conditions. For $x = x_0 < 0$ when $t = 0$ the solution of x is

$$x = x_0 e^t$$

and as $t \to \infty$, $x \to -\infty$ (because $x_0 < 0$).

For $x = x_0 > 0$ when $t = 0$ the solution of x is

$$x = x_0 e^t$$

and as $t \to \infty$, $x \to +\infty$ (because $x_0 < 0$). The phase portrait for this velocity field is shown in Fig 2.8.

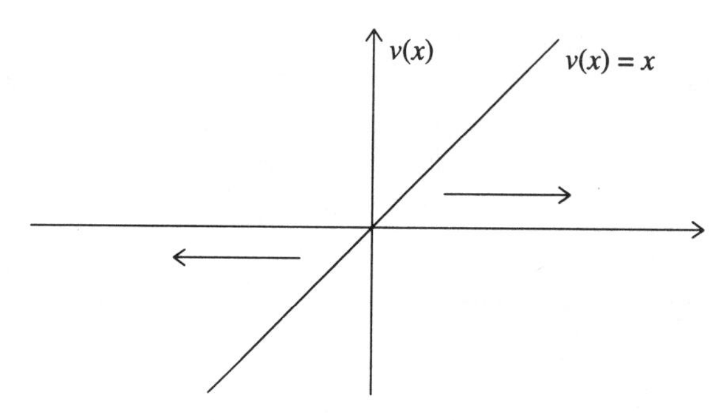

Fig 2.8 Phase portrait for $v(x) = x$

From the phase portrait we can deduce that there is a fixed point at the origin. If the initial point is at $x > 0$, say $x_0 = 1$, then the point moves away from the origin to infinity with increasing speed. If the initial point is at $x < 0$, say $x_0 = -1$, then the point moves away from the origin to negative infinity with increasing speed. The extra information that the analytical solution provides is timescales.

Exercises 2.2

1. Draw the phase portraits and find the fixed points and invariant sets of the systems with the following velocity functions. Here a and b are any pair of numbers, with $b > a$.

   (i) $v(x) = (x - a)(x - b)$

   (ii) $v(x) = (a - x)(x - b)$

   Give a qualitative description of the motion of the system.

2. Draw the phase portrait and find the fixed points and invariant sets of the system with the velocity function $v(x) = \sin x$.

3. Draw the phase diagram of the system

$$\frac{dx}{dt} = b(x - x_f)$$

   where b is a constant. Find the fixed points and describe the behaviour of the system qualitatively. (Consider the cases $b > 0$, $b < 0$ separately.) Solve the differential equation to validate your description.

4. Draw the phase diagram of the dynamical system

$$\frac{dx}{dt} = 2x^4 - 3x^3 - 4x^2 + 3x + 2$$

   Find the fixed points and the invariant sets and describe qualitatively the motion of the system.

2.3 Attractors and repellors

The fixed points are clearly important in describing the motion of the system. If the initial state is at a fixed point then the system does not change. However, if the system is displaced slightly from a fixed point its motion depends on the velocity field near the fixed point.

There are essentially four types of fixed points shown in Fig 2.9.

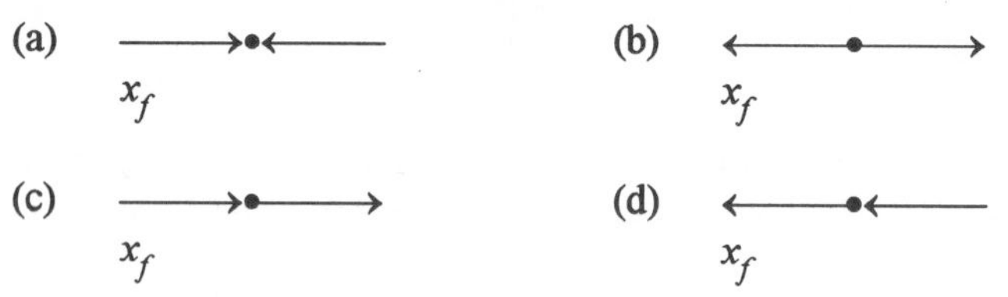

(a) $\longrightarrow\bullet\longleftarrow$ x_f (b) $\longleftarrow\bullet\longrightarrow$ x_f

(c) $\longrightarrow\bullet\longrightarrow$ x_f (d) $\longleftarrow\bullet\longleftarrow$ x_f

Fig 2.9 The four possible phase portraits associated with a fixed point at $x = x_f$

In part (a), if the system is given a small displacement from the fixed point $x = x_f$ then it will tend to return to the fixed point because the direction field is towards $x = x_f$ on each side. The fixed point is said to be **stable** and is described as an **attractor**.

In part (b), if the system is given a small displacement from the fixed point $x = x_f$ then it will tend to move away from the fixed point. The fixed point is said to be **unstable** and is described as a **repellor**. For parts (c) and (d) the system is neither stable nor unstable. From one side the system appears to be stable and from the other side it appears to be unstable. The fixed point is said to be semi-stable and is described as a **shunt** in each of these cases.

TUTORIAL PROBLEM 2.4

Draw the phase portraits for the linear system

$$\frac{dx}{dt} = a + bx \qquad\qquad b \neq 0$$

for the two cases $b > 0$ and $b < 0$. In each case classify the fixed point as an attractor or a repellor.

In Chapter 1 we found the general solution of the linear system to be

$$x = x_f + (x_0 - x_f)e^{bt}$$

where $x_f = -\dfrac{a}{b}$ is the fixed point and $x = x_0$ is the initial phase point $x(0)$. This solution can be used to confirm the behaviour of the system:

for $b > 0$ $\quad x \to \infty$ as $t \to \infty$ and $x = x_f$ is a repellor

for $b < 0$ $\quad x \to x_f$ as $t \to \infty$ and $x = x_f$ is an attractor.

This linear system turns out to be very important in describing the behaviour of non-linear systems near a fixed point. Consider the general velocity field $v(x)$ and let $x = x_f$ be a fixed point of the system so that $v(x_f) = 0$. Suppose we take the first order Taylor polynomial expansion of $v(x)$ about x_f then,

$$v(x) = v(x_f) + v'(x_f)(x - x_f) + 0((x - x_f)^2)$$

since x_f is a fixed point $v(x_f) = 0$. Assume that $v'(x_f) \neq 0$. Such a fixed point is said to be **simple**. Then to first order in $(x - x_f)$ we have

$$v(x) = v'(x_f)(x - x_f)$$

If we compare this with the linear system $v(x) = a + bx$ then we can deduce immediately that if $v'(x_f) < 0$ the fixed point is stable, and if $v'(x_f) > 0$ the fixed point is unstable.

This provides a powerful method of analysing many non-linear systems. The Taylor polynomial near a simple fixed point is called the **linearized approximation** of the velocity field in the vicinity of the fixed point. Such a fixed point is stable if all sufficiently close neighbouring phase states approach x_f as $t \to \infty$. A fixed point is unstable if all sufficiently close neighbouring phase states leave x_f as $t \to \infty$.

TUTORIAL PROBLEM 2.5

Draw the phase portraits for the quadratic system

$$\frac{dx}{dt} = b(x - x_f)^2 \qquad b \neq 0$$

for the two cases $b > 0$ and $b < 0$. Classify the fixed point in each case. Is the fixed point simple? Show that the motion of a nearby point is not exponential.

The quadratic velocity field of Tutorial Problem 2.5 shows an important feature called **terminating motion**. The general solution of the system

$$\frac{dx}{dt} = b(x - x_f)^2 \qquad b > 0$$

with initial condition $x(0) = x_0$ is

$$x = x_f + \frac{x_0 - x_f}{1 - b(x_0 - x_f)t} \qquad t < \frac{1}{b(x_0 - x_f)}$$

This solution shows that as $t \to \dfrac{1}{b(x_0 - x_f)}$, $x \to \infty$ and we say that the motion **terminates** at the critical time $\tau = \dfrac{1}{b(x_0 - x_f)}$. This is called the **terminating time** for a finite time interval. So for the quadratic system, the solution only exists for $0 < t < \tau$, i.e. for a finite time interval. In real situations the model would break down before the terminating time is reached.

TUTORIAL PROBLEM 2.6

Consider the system with velocity field

$$v(x) = (x - x_f)^n$$

and initial condition $x(0) = x_0$, $x_0 \neq 0$. Show that for all positive integers $n \geq 2$ the motion is terminating. Find the terminating time.

Exercises 2.3

1. Show that if $v(x)$ is a decreasing function of x near the fixed point x_f then x_f is stable, while if $v(x)$ is increasing near x_f then x_f is unstable.

2. Show that all the fixed points of the dynamical system with velocity function $v(x) = x(x^2 - 1)$ are simple, and classify them as stable or unstable.

3. Find and classify the fixed points of the systems with the following velocity functions, where $b > a$:

   (i) $v(x) = (x - a)(x - b)$

   (ii) $v(x) = (a - x)(x - b)$

4. Consider an autonomous system which has exactly two simple fixed points. Can both fixed points be stable? Can both fixed points be unstable? How many essentially different phase diagrams can there be for such a system?

5. Suppose that x_f is a fixed point of a system with velocity function $v(x)$, that $v'(x_f) = 0$ but that $v''(x_f) \neq 0$. Discuss the nature of the fixed point x_f.

   Illustrate your answer by choosing a velocity function $v(x)$ with the required conditions.

6. Draw the phase portrait for the velocity function $v(x) = dx^3$, where d is a non-zero constant. Show that a fixed point which is not simple may nevertheless be stable or unstable. Show that the motion of a phase point near the fixed point is not exponential in this case, but that nevertheless, when the fixed point is stable no other point can reach it in a finite time.

7. Find the fixed points of the following systems. Investigate their behaviour near the fixed points using Taylor polynomial expansions and hence classify them as simple, stable, unstable or shunt.

(i) $\dfrac{dx}{dt} = x + 5$ (ii) $\dfrac{dx}{dt} = 1 + x^2$

(iii) $\dfrac{dx}{dt} = x^4 - x^3 - 2x^2$ (iv) $\dfrac{dx}{dt} = 2x^3 - 3x^2 + 1$

(v) $\dfrac{dx}{dt} = \sinh(x^2)$

8. Consider the non-linear system

$$\frac{dx}{dt} = x^3 - ax - b$$

Find a relationship between a and b so that (i) the phase portrait consists of a single repellor, and (ii) the phase portrait consists of two repellors separated by an attractor.

9. In an experiment to model the growth of sweet pea plants it is found that the height of a plant, h cm, satisfies the logistic differential equation

$$\frac{dh}{dt} = 0.65h\left(1 - \frac{h}{192}\right)$$

where t is the time in weeks. Find and classify the fixed points. Explain the physical significance of the fixed points in this example.

10. The logistic differential equation

$$\frac{dP}{dt} = k(M - P)(P - m)$$

is used to model the population of whales, P, as a function of time t, where k, M and m are constants. If the number of whales falls below a minimum level the species will become extinct. The population is also limited to a maximum number by the carrying capacity of the environment.

(i) Draw the phase portrait for this model and deduce the physical meaning of the constants m and M.

(ii) Describe the change to the system if the initial whale population P_0 satisfies (a) $P_0 < m$, (b) $m < P_0 < M$, and (c) $M < P_0$.

(iii) Assuming that the model reasonably estimates the whale population, what implications are suggested for fishing? What controls would you suggest?

2.4 Natural boundaries

In some problems the dependent variable is restricted in the meaningful values it can take. For example in modelling the species of a population the variable P representing the number of the species must take non-negative values. A negative population has no meaning. The phase space is then restricted to the non-negative real line $P \geq 0$. The phase point $P = 0$ is called a **natural boundary** and the phase space is restricted to the interval $[0,\infty]$.

In some problems natural boundaries occur because of mathematical restrictions. Take the velocity field

$$v(x) = \sqrt{1-x^2}$$

This has real values on the interval $[-1,1]$ so the phase points $x = -1$ and $x = 1$ are natural boundaries. Although the velocity field is well defined at the natural boundaries, it does not have a Taylor polynomial expansion there so we can expect peculiar behaviour near each natural boundary. The phase portrait in Fig 2.10 shows that for any initial value $x = x_0$ the solution moves towards the phase point $x = 1$, but the behaviour will be different from a stable fixed point or a shunt.

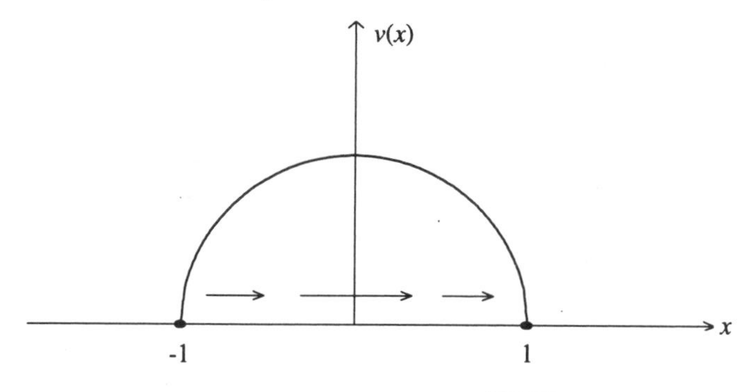

Fig 2.10 Phase portrait of $v(x) = \sqrt{1-x^2}$

Consider the motion near to the fixed point $x = 1$. We can write $v(x)$ as a product of functions

$$v(x) = \sqrt{1-x}\sqrt{1+x}$$

Now $\sqrt{1+x}$ is well behaved near $x = 1$ and it can be written as a Taylor polynomial expansion in the form

$$\sqrt{1+x} = \sqrt{2} - \frac{1}{2\sqrt{2}}(1-x) + 0\left((1-x)^2\right)$$

With this expansion $v(x)$ can be written as

$$v(x) = \sqrt{2}\,(1-x)^{\frac{1}{2}} - \frac{1}{2\sqrt{2}}(1-x)^{\frac{3}{2}} + 0\left((1-x)^{\frac{5}{2}}\right)$$

If the initial condition x_0 is close to the natural boundary $x = 1$ then

$$\frac{dx}{dt} = v(x) \approx \sqrt{2}\,(1-x)^{\frac{1}{2}}$$

Using the variables separable method of solution,

$$\int_{x_0}^{x} (1-x)^{-\frac{1}{2}}\,dx = \int_0^t \sqrt{2}\,dt$$

$$-2(1-x)^{\frac{1}{2}} + 2(1-x_0)^{\frac{1}{2}} = \sqrt{2}\,t$$

Solving for x we have

$$x = 1 - \left((1-x_0)^{\frac{1}{2}} - \frac{t}{\sqrt{2}}\right)^2$$

The system approaches the natural boundary $x = 1$ in a time $t = \sqrt{2(1-x_0)}$. Beyond this time no motion provided by the model $v(x) = \sqrt{1-x^2}$ is possible. The system approaches the boundary in a finite time and motion terminates there.

TUTORIAL PROBLEM 2.7

Investigate the motion of the velocity field

$$v(x) = \sqrt{1-x^2}$$

near the natural boundary $x = -1$. Show that the equation of motion may be solved analytically using the substitution $x = \sin y$. Confirm that the motion terminates at the natural boundary $x = 1$ for all initial positions.

Exercises 2.4

1. (i) Consider the first order autonomous system with the velocity function

$$v(x) = x^3 - 2x^2 + x$$

where x is allowed to take a real value.

(a) Locate the fixed points and elementary invariant sets of this system, and sketch its phase diagram. For each fixed point, state whether or not it is simple and whether it is stable, unstable or neither.

(b) Describe the motions for $t > 0$ of a phase point initially at $x = 1.5$, giving as much detail as you can, and in particular stating whether or not the motion terminates.

(c) The system is perturbed by the addition to its velocity function of a small constant term (which is non-zero but may be either positive or negative). Discuss how the structure of the flow is changed, if at all, by the perturbation.

(ii) Consider the system with velocity function $\sqrt{v(x)}$, where x is now restricted to the range $0 \leq x \leq 1$, and where $v(x)$ is the function of part (i) (and for any positive k, $\sqrt{k}$ is the *positive* number whose square is k). Investigate the motions of this system, being careful to state where any natural boundaries are located, and paying particular attention to motion near natural boundaries and fixed points.

2. Investigate the first order autonomous system with velocity function

$$v(x) = 3x^{\frac{2}{3}}$$

(i) Are there any natural boundaries?
(ii) Is there terminating motion?

3. Show that the motion of the system

$$\frac{dx}{dt} = v(x) = x^{\alpha} \qquad x > 0$$

terminates for any α greater than 1.

4. Show that, if $f(x)$ is a well-behaved function such that $f(b) = 0$ and $f'(b) < 0$, so that the Taylor expansion of $f(x)$ about b may be written as

$$f(x) = -a^2(x - b) + 0(x - b^2)$$

for some real, non-zero constant a, then the motion defined by the velocity field

$$v(x) = \sqrt{f(x)}$$

terminates at $x = b$.

5. Describe the motion with velocity field

$$w(x) = \sqrt{g(x)}$$

where $g(x)$ is a polynomial with N distinct real roots.

2.5 Case study: Population growth

This section shows how the ideas of this chapter can be used in modelling real situations.

The need to model the population of different species is important, but the process is not easy. There are many features to take into account, certainly too many for a simple model. For example, consider the problem of modelling the population change of a rare species of bird on an island. A typical list of features might look like this:

1. initial number of birds
2. size of the island
3. availability of food
4. number of males and females
5. age distribution of birds
6. life-span of birds
7. types of predators
8. no. of eggs per female
9. population of predators
10. diseases
11. proneness of birds to fight
12. family size
13. role of unattached males
14. will birds emigrate
15. immigration of birds
16. human exploitation.

This list can be refined and sorted into four main headings:

- features affecting births
- features affecting deaths
- features affecting food
- emigration and immigration.

The features affecting food such as the size of the island, availability of food and family size will clearly have a bearing on births and deaths.

Now we take what might seem a big step. Consider the four headings. They suggest the following equation in words:

$$\begin{pmatrix}\text{change in population}\\ \text{of birds during}\\ \text{a period of time}\end{pmatrix} = \begin{pmatrix}\text{number of}\\ \text{births}\end{pmatrix} - \begin{pmatrix}\text{number of}\\ \text{deaths}\end{pmatrix} + (\text{immigrants}) - (\text{emigrants})$$

The features affecting food will affect the numbers of births and deaths and the ability of the island to sustain a large number of birds. Since the species of birds is rare it is likely that there will be no immigrants to the island in the early years. We will make a further simplification that there is no net emigration from the island, i.e. all those birds that leave the island return the following year.

Suppose that $P(t)$ denotes the size of the population t years after the birds were introduced onto the island, so that $P(t+h) - P(t)$ is the increase in the population during a time interval h. Let B and D be the number of births and deaths respectively during time interval h, and then the word model can now be written in symbols as

$$P(t+h) - P(t) = B - D$$

The next step is to express the number of births and deaths in terms of population size and/or time. There could be many ways forward from here, but we will keep it simple by making the following assumptions:

1. the number of births is proportional to the size of the population;
2. the number of deaths is proportional to the size of the population;
3. the numbers of births and deaths are proportional to the time interval h.

In symbols we have

$$B = bhP \quad \text{and} \quad D = dhP$$

where b and d are constants, called the proportionate birth rate and the proportionate death rate respectively. Hence the simple mathematical model for the bird population is

$$P(t+h) - P(t) = Ph(b - d)$$

Dividing by h and letting $h \to 0$ we can formulate a calculus model implying continuous changes

$$\frac{\mathrm{d}P}{\mathrm{d}t} = P(b-d) = Pg(P)$$

where $g(P)$ is the proportionate growth rate of the bird population. The jump to a continuous variable for P is an assumption that could be seriously criticised. Although deaths might be continuous, births are likely to be at particular times of the year only.

Now we can investigate different growth rate models. The simplest assumption is that the proportionate birth and death rates are constants so that $g(P) = c$. The differential equation model is then linear

$$\frac{\mathrm{d}P}{\mathrm{d}t} = cP$$

If $c < 0$ then the population dies out, P is a stable fixed point. This is not very good news for the bird population! If $c > 0$ then the population increases exponentially without bound, P is an unstable fixed point. This is unlikely to happen because there are factors which limit population size such as the availability of space and food.

If $c = 0$ the population does not change.

A more common model for $g(P)$ is a decreasing function of P.

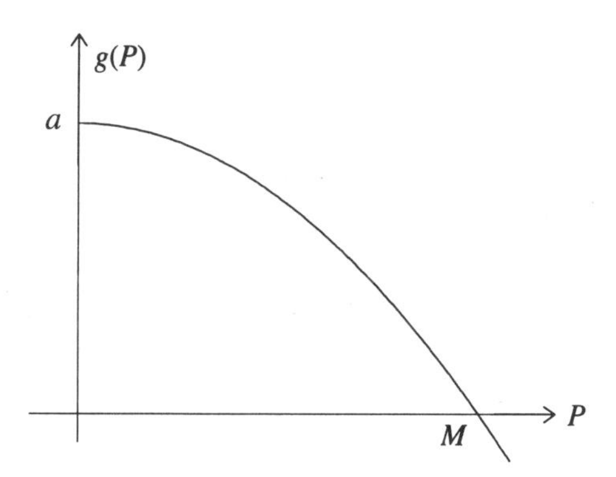

Fig 2.11 Proportionate growth rate for real populations

When P is small, $g(P) \sim a > 0$ and the differential equation model is

$$\frac{\mathrm{d}P}{\mathrm{d}t} \sim aP$$

The population initially grows exponentially. As the population increases, the proportionate growth rate tends to zero and then does not change.
When $P = M$, $g(P) = g(M) = 0$ and the differential equation model is

$$\frac{\mathrm{d}P}{\mathrm{d}t} \sim 0$$

i.e. the population P remains constant at $P = M$.

In Fig 2.11 the initial growth rate is a and when $P = M$ the growth rate is zero, so $g(0) = a$ and $g(M) = 0$. The simplest model for $g(P)$ is a linear function of P given by

$$g(P) = a\left(1 - \frac{P}{M}\right)$$

In this case the differential equation is called **the logistic model** and has velocity field

$$v(P) = aP\left(1 - \frac{P}{M}\right)$$

The next tutorial problem investigates the logistic model using the techniques of this chapter.

TUTORIAL PROBLEM 2.8

Consider a model for proportionate growth rate as a linearly decreasing function

$$g(P) = a\left(1 - \frac{P}{M}\right)$$

where $g(0) = a$ = constant.

(i) Find and classify the fixed points for the logistic model. Describe qualitatively the change in population with time for the cases (a) $P(0) = P_0 < M$, and (b) $P(0) = P_0 > M$.

(ii) Without actually solving the equation, show that the solution with initial condition $P = P_0$ at $t = 0$ behaves approximately like

$$P(t) = P_0 e^{at}$$

for small P_0 and small t.

(iii) Find an exponential approximation to the solution for P_0 close to M. (Hint: choose P_0 close to M and find a Taylor polynomial for $v(P)$.)

(iv) The full analytical solution is

$$P(t) = \frac{M}{1 + \left(\frac{M}{P_0} - 1\right)e^{-at}}$$

Use this expression to confirm your answers to parts (ii) and (iii).

The human species has learnt how to make use of the environment and nature through fishing, farming, forestry, biological pest control and so on. The exploitation of the world's natural resources is fraught with dangers unless there are controls and appropriate management to avoid over-harvesting. Over-fishing of the sea has been a particular problem since the 1950s and we now have such low fish stocks in some parts of the oceans that some species may be beyond recovery.

We can use our simple modelling ideas of this section to investigate the exploitation of a species. Suppose that the population of a species of fish which inhabits parts of a sea grows logistically according to the model

$$\frac{dP}{dt} = aP\left(1 - \frac{P}{M}\right)$$

where M is the carrying capacity of the sea.

To model the effects of fishing we assume that the catch is proportional to the population of fish present, so that the logistic model becomes

$$\frac{\mathrm{d}P}{\mathrm{d}t} = aP\left(1-\frac{P}{M}\right) - eP$$

where e is a constant, often called the **effort**. This fishing technique assumes no knowledge of where the fish are located and fishermen just drag their nets through the water in a 'blindfold' manner.

There are fixed points at $P_1 = 0$ and $P_2 = \frac{M(a-e)}{a}$. The phase portrait, shown in Fig 2.12, has two forms depending on whether $a > e$ or $a < e$.

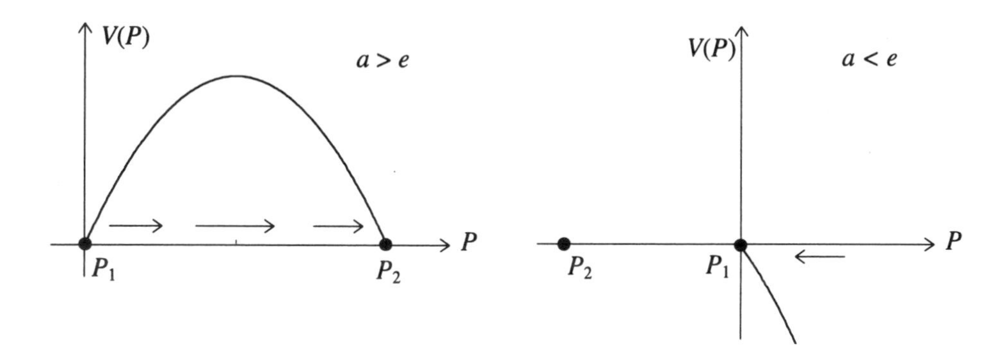

Fig 2.12 Phase portrait for blind fishing

There is a natural boundary at $P = 0$, so in each case we have only drawn the phase portrait for $P \geq 0$.

- If $e < a$ then there is a stable population of fish stocks given by $P = P_2 = \frac{M(a-e)}{a}$.

- If $e > a$ the population of fish will decrease to zero. This is clearly what has been occurring in parts of the North Sea where cod fishing during the post-war period has seen stocks seriously depleted.

This simple model shows how the techniques of this chapter can be used to analyse real situations. In the next tutorial problem you investigate a fishing strategy of removing fish at a constant harvesting rate from the population.

TUTORIAL PROBLEM 2.9

An alternative model for fishing assumes that the fishermen know where the fish are located and decide how many fish to catch in a given amount of time. This leads to the **constant harvesting model**

$$\frac{dP}{dt} = aP\left(1 - \frac{P}{M}\right) - c$$

where c is a constant called the **harvesting rate**.

(i) Find and classify the fixed points P_1 and P_2 of the constant harvesting population growth model.

(ii) Describe the future of the species if the current population is P_c for the three cases (a) $0 < P_c < P_1$, (b) $P_1 < P_c < P_2$, (c) $P_2 < P_c$.

Exercises 2.5

1. Various modifications of the logistic model have been proposed. One particular model is the Gompertz equation

$$\frac{dP}{dt} = rP(a - \ln P)$$

where r and a are constants.

(i) Find and classify the fixed points of this model.

(ii) Describe qualitatively the change of population with time for this model.

(iii) From your description in part (ii), sketch possible solution curves for P against t.

(iv) What is the limiting population as $t \to \infty$?

2. Consider a general population growth model with constant catch exploitation

$$\frac{dP}{dt} = P(g(P) - e)$$

Analyse the possible different outcomes for this model.

3. A farmer in Devon is planning to issue deer hunting permits on his estate. The farmer knows that if the deer population falls below a certain level m the deer will become extinct. He also knows that his estate has a maximum carrying capacity M, so that if the population goes above m then it will increase to M. A model for population growth of the deer in the park is found to be

$$\frac{dP}{dt} = kP(M-P)(P-m)$$

where k is constant.

(i) Draw the phase portrait for this population growth model.

(ii) Investigate and describe qualitatively the change in population with time for the current deer population P_c in the three cases (a) $P_c < m$, (b) $m < P_c < M$, (c) $P_c > M$

Further exercises

1. Sketch the phase portrait for each of the following systems. Find and classify the fixed points in each case. Describe qualitatively the motion of the system.

(i) $\frac{dx}{dt} = bx$ (ii) $\frac{dx}{dt} = 5x - x^2$

(iii) $\frac{dx}{dt} = 4x - x^3$ (iv) $\frac{dx}{dt} = 6x^4 + 7x^3 - 9x^2 - 7x + 3$

(v) $\frac{dx}{dt} = \cos x$ (vi) $\frac{dx}{dt} = 3x^{\frac{5}{2}}$

2. Sketch the phase portrait for the system with velocity field

$$v(x) = x^3 + \alpha x$$

where α is a constant parameter.

(i) Describe how the number of fixed points, and their nature, depend on the value of α.

(ii) For what value(s) of α is there terminating motion for any starting value and find the corresponding terminating time.

(iii) Does the system have any natural boundaries?

3. Explain the term 'terminating motion'. The following figures show the phase portrait of two first order non-linear systems.

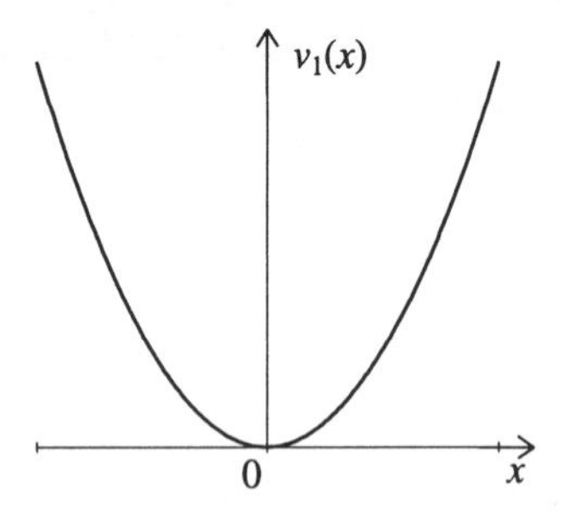

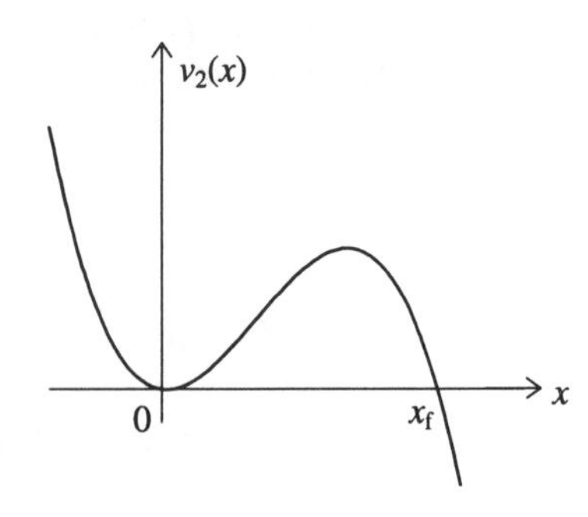

Fig 2.13 Two non-linear systems

$v_1(x)$ and $v_2(x)$ have non-simple fixed points at the origin. The Taylor polynomial approximation of $v_1(x)$ and $v_2(x)$ close to the origin is of the form $x^2 + 0(x^4)$. Explain why $v_1(x)$ is terminating motion and $v_2(x)$ is not. For $v_1(x)$ find the terminating time.

4. A substance C is formed in a chemical reaction between substances A and B. In the reaction each gram of C is produced by the combination of p grams of A and $q = 1 - p$ grams of B. The rate of formation of C at any instant in time, t, is equal to the product of the masses of A and B that remain uncombined at that instant.

   If a grams of A and b grams of B are bought together at $t = 0$, show that the differential equation governing the mass, $x(t)$, of c present at time $t > 0$ is

$$\frac{dx}{dt} = (a - px)(b - qx)$$

   Assuming $\frac{a}{p} > \frac{b}{q}$, construct the phase diagram for this equation. What is the maximum amount C that can possibly be produced in this experiment?

5. A ball bearing of mass m falls vertically through a tall cylinder of oil. The equation of motion of the ball is

$$m\frac{dv}{dt} = 9.8m - kv^2$$

   where v is the velocity and k is a constant.

   (i) Draw a phase portrait for this system and give a qualitative description of the motion of the ball.

(ii) Find an analytical solution of the equation of motion and confirm your qualitative description given in part (i).

6. A tank made of porous material has a square base of side 2 metres and vertical sides. The tank contains water which seeps out through the base and sides at a rate proportional to the total area in contact with the water. When the depth is 3 metres, it is observed to be falling at a rate of 0.2 metres per hour.

(i) Formulate a differential equation to model this situation.

(ii) Sketch the phase portrait for your model, identifying any natural boundaries and fixed points.

(iii) Describe the motion of the system qualitatively.

7. The rate at which a substance is deposited from a solution to form a crystal is proportional to the product of the mass already deposited, m, and the mass remaining in the solution. The solution originally contained 20 grams of the substance.

(i) Formulate a differential equation model for m.

(ii) Sketch the phase portrait for your model, identifying any natural boundaries and fixed points.

(iii) Describe the motion of the system qualitatively for different initial conditions.

8. A system has velocity field

$$v(x) = \sqrt{2-x}$$

(i) Sketch the phase portrait for this system, identifying any natural boundaries and fixed points.

(ii) Describe the motion of the system qualitatively.

(iii) Find the terminating time for a phase point that is initially at $x_0 = 1.5$.

3 •Second Order Continuous Autonomous Systems

In Chapter 2 we investigated the motion of first order systems which are modelled by one first order differential equation. The phase portrait for a first order autonomous system consists of arrows drawn parallel to the x-axis (i.e. the axis of the dependent variable). We shall see that second order autonomous systems show more complicated and (probably) more interesting behaviour. The phase space consists of a phase plane and a good understanding of the motion of a system can be obtained from direction field type diagrams.

3.1 Autonomous second order systems

A second order system is defined by two dependent variables x_1 and x_2 modelled by two simultaneous first order differential equations

$$\frac{dx_1}{dt} = v_1(x_1, x_2, t)$$

$$\frac{dx_2}{dt} = v_2(x_1, x_2, t)$$

The functions v_1 and v_2 are the components of a vector called the **velocity field** $\mathbf{v} = (v_1,v_2)$. The value of the velocity field at given values of x_1, x_2 and t is called the **phase velocity**. If $\mathbf{x} = (x_1,x_2)$ is a vector with components x_1 and x_2 then the system can be written in vector form as

$$\frac{d\mathbf{x}}{dt} = \mathbf{v}(\mathbf{x}, t)$$

The system is autonomous if $\mathbf{v}$ does not depend explicitly on t so that

$$v_1 = v_1(x_1,x_2) \quad \text{and} \quad v_2 = v_2(x_1,x_2)$$

You may be familiar with second order systems from mechanics which are modelled by second order differential equations. The following example shows that this is consistent with the definition given above.

Example 1

By choosing new variables $x_1 = x$ and $x_2 = \frac{dx}{dt}$ show that the second order differential equation

$$\frac{d^2x}{dt^2} - \frac{3dx}{dt} + 4x = \sin t$$

can be written as two first order differential equations.

Solution

Let $x_1 = x$ and $x_2 = \frac{dx}{dt}$; then $\frac{dx_2}{dt} = \frac{d^2x}{dt^2}$. Replace x by x_1, $\frac{dx}{dt}$ by x_2 and $\frac{d^2x}{dt^2}$ by $\frac{d^2x_2}{dt}$ in the second order differential equation to obtain

$$\frac{dx_2}{dt} - 3x_2 + 4x_1 = \sin t$$

i.e. $$\frac{dx_2}{dt} = -4x_1 + 3x_2 + \sin t$$

This equation, together with $\frac{dx_1}{dt} = x_2$, gives a pair of simultaneous first order differential equations. A system described by these equations would be non-autonomous because of $\sin t$. There is no unique way of converting a second order equation into a pair of simultaneous first order equations.

TUTORIAL PROBLEM 3.1

Write the second order differential equation

$$\frac{d^2x}{dt^2} + \frac{dx}{dt} + x - x^3 = 0$$

as pairs of first order equations using the following changes of variable:

(i) $x_1 = x$ and $x_2 = \frac{dx}{dt}$

(ii) $x_1 = x$ and $x_2 = \frac{dx}{dt} + x$.

For problems in mechanics the two dependent variables are usually chosen to have physical significance. The variable $x_1 = x$ is the position of the object from a fixed origin and $x_2 = m\dfrac{dx}{dt}$ is the linear momentum of the object of mass m. Newton's second law provides a model for most situations

$$m\frac{d^2x}{dt^2} = F\left(x, \frac{dx}{dt}, t\right)$$

where F is the force which could depend on position, velocity and time.

Differentiating x_2 we get $\dfrac{dx_2}{dt} = m\dfrac{d^2x}{dt^2}$.

Substituting for x $(= x_1)$, $\dfrac{dx}{dt}\left(= \dfrac{x_2}{m}\right)$ and $m\dfrac{d^2x}{dt^2}\left(= \dfrac{dx_2}{dt}\right)$ we have a pair of simultaneous first order equations

$$\frac{dx_2}{dt} = F\left(x_1, \frac{x_2}{m}, t\right)$$

$$\frac{dx_1}{dt} = \frac{x_2}{m}$$

The velocity field for this system is $\mathbf{v} = \left(\dfrac{x_2}{m}, F\left(x_1, \dfrac{x_2}{m}, t\right)\right)$. Note the potential confusion with the language of mechanics and non-linear systems. The vector **v** is not the velocity vector in the normal sense of the word, so care is needed when applying the methods of non-linear systems to problems in mechanics.

Exercises 3.1

1. Which of the following velocity fields are autonomous?

(i) $\dfrac{dx_1}{dt} = x_2, \qquad \dfrac{dx_2}{dt} = -x_1$

(ii) $\dfrac{dx_1}{dt} = x_1^2 - x_2 t, \qquad \dfrac{dx_2}{dt} = -x_2 + x_1$

(iii) $\dfrac{dx_1}{dt} = x_1 + x_2, \qquad \dfrac{dx_2}{dt} = \sin t$

2. Write the following second order differential equations as pairs of simultaneous first order equations:

   (i) $\dfrac{d^2x}{dt^2} + \lambda \sin x = 0$, where λ is a constant.

   (ii) $\dfrac{d^2x}{dt^2} + 2a\dfrac{dx}{dt} + b = 0$, where a and b are constants.

   (iii) $\dfrac{d^2x}{dt^2} + \lambda x = k \sin \omega t$ where λ, k and ω are constants.

3. The equation of motion of a stone of mass m falling freely in a vertical direction is

$$\frac{d^2x}{dt^2} = -g$$

   where x is the vertical height and g is the acceleration due to gravity. What is the velocity field for the second order system corresponding to this equation?

4. A ball bearing of mass 0.1 kg falls vertically through a tall cylinder of oil. The equation of motion of the ball is

$$0.1\frac{d^2x}{dt^2} = 0.98 - 0.05v^2$$

   where v is the speed of the ball and x is the position of the ball from the top of the cylinder. What is the velocity field for the second order system corresponding to this equation?

5. Show that the substitution $x_1 = x$ and $x_2 = t$ changes the first order non-autonomous system

$$\frac{dx}{dt} = 2x - \cos 3t$$

   into a second order autonomous system. Write down the velocity field. Hence show that any first order non-autonomous system

$$\frac{dx}{dt} = v(x, t)$$

   can be changed into a second order autonomous system. Write down the velocity field.

3.2 Constant coefficient equations

The link between a pair of simultaneous first order differential equations and an associated second order differential equation provides a simple method of solution for constant coefficient autonomous systems.

Consider the second order system modelled by

$$\frac{dx_1}{dt} = 2x_1 + 4x_2 \tag{3.1}$$

$$\frac{dx_2}{dt} = x_1 - x_2 \tag{3.2}$$

We start by making x_1 the subject of the equation (3.2):

$$x_1 = \frac{dx_2}{dt} + x_2 \tag{3.3}$$

Differentiating with respect to t gives

$$\frac{dx_1}{dt} = \frac{d^2 x_2}{dt^2} + \frac{dx_2}{dt}$$

Substitute for x_1 and $\frac{dx_1}{dt}$ into the equation (3.1) to give

$$\frac{d^2 x_2}{dt^2} + \frac{dx_2}{dt} = 2\left(\frac{dx_2}{dt} + x_2\right) + 4x_2$$

This simplifies to the second order differential equation with constant coefficients

$$\frac{d^2 x_2}{dt^2} - \frac{dx_2}{dt} - 6x_2 = 0$$

The general solution of this equation is

$$x_2 = Ae^{3t} + Be^{-2t}$$

This parametric equation in x_2 is part of the general solution of the original pair of simultaneous first order equations. Substituting into equation (3.3) gives the general solution for x_1, the other part of the general solution of the original system:

$$x_1 = 4Ae^{3t} - Be^{-2t}$$

Figure 3.1 shows graphs of $x_1(t)$ and $x_2(t)$ for $A = 1$ and $B = 1$ which correspond to initial conditions $x_1 = 3$ and $x_2 = 2$ when $t = 0$.

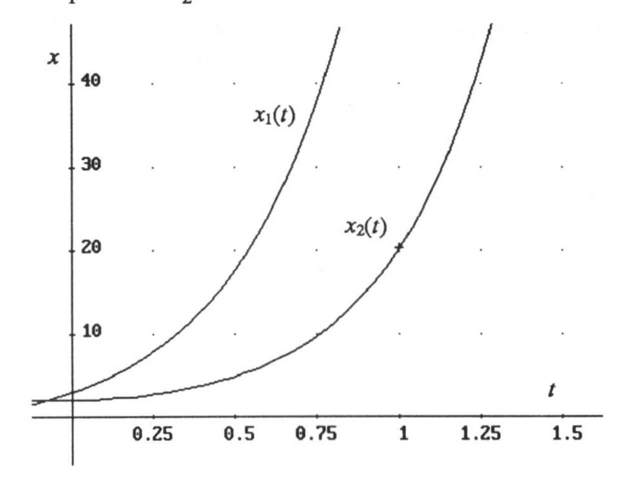

Fig 3.1 The particular solution of the system described by equations (3.1) and (3.2)

We can see that as t increases, x_1 and x_2 both grow indefinitely.

TUTORIAL PROBLEM 3.2

A system is modelled by the pair of simultaneous differential equations given by

$$\frac{dx_1}{dt} = -x_1 + x_2 - 1$$

$$\frac{dx_2}{dt} = -x_1 - x_2 + 3$$

and the initial conditions $x_1 = 0$ and $x_2 = 3$ when $t = 0$.

(i) Find expressions for x_1 and x_2 in terms of t.

(ii) Describe what happens as $t \to \infty$.

Exercises 3.2

1. For each of the following pairs of simultaneous first order differential equations.

   (a) Solve the equations to find expressions for x_1 and x_2 in terms of t.

   (b) Find the particular solutions for which $x_1 = 1$ and $x_2 = 2$ at $t = 0$.

(c) Describe the long term behaviour of each of the following systems:

(i) $\dfrac{dx_1}{dt} = 3x_1 - x_2, \qquad \dfrac{dx_2}{dt} = 2x_1$

(ii) $\dfrac{dx_1}{dt} = x_1 + x_2, \qquad \dfrac{dx_2}{dt} = x_1 - x_2$

(iii) $\dfrac{dx_1}{dt} = x_1 + 2x_2 - 3, \qquad \dfrac{dx_2}{dt} = -3x_1 + x_2 + 2$

(iv) $\dfrac{dx_1}{dt} = 2x_1 + 3x_2, \qquad \dfrac{dx_2}{dt} = 3x_1 + 2x_2$

2. A population of cells consisting of a mixture of 2-chromosome and 4-chromosome cells is described approximately by

$$\frac{dT}{dt} = (a-b)T, \qquad \frac{dF}{dt} = bT + cF$$

where T is the number of 2-chromosome cells and F is the number of 4-chromosome cells; T and F are both non-negative; a, b and c are constants with $a \neq b$ and $c \neq 0$.

Show that, whatever the values of a, b and c, in the long term the proportion of 2-chromosome cells in the cell population, defined by $R = T/(F + T)$, tends to a constant value independent of the initial conditions.

Find the conditions on the values of a, b and c which ensure that this limiting value of the proportion is non-zero, and find the limit in this case.

3. A simple model of a predator–prey system is described by the pair of linear differential equations

$$\frac{dx}{dt} = x + y \qquad \frac{dy}{dt} = -x + y$$

where x is the number of predators and y the number of prey. If the initial population of both species is 500 find the particular solutions for x and y. Describe what happens to the population of predator and prey as time increases.

3.3 Phase curves and fixed points

The problems in Section 3.2 have solutions in which the dependent variables x_1 and x_2 vary with t. For example, the particular solution of Tutorial Problem 3.2 is

$$x_1 = -\,e^{-t}\cos t + e^{-t}\sin t + 1$$
$$x_2 = e^{-t}\sin t + e^{-t}\cos t + 2 \qquad (3.4)$$

As t increases x and y approach the values 1 and 2 respectively. We deduce that in the long term the system approaches the point (1, 2).

The pair of differential equations for this system is

$$\frac{dx_1}{dt} = -x_1 + x_2 - 1$$

$$\frac{dx_2}{dt} = -x_1 - x_2 + 3$$

At the phase point (1,2),

$$\frac{dx_1}{dt} = -1 + 2 - 1 = 0$$

$$\frac{dx_2}{dt} = -1 - 2 + 3 = 0$$

Hence the velocity field is zero at the point (1,2). Such a point is called **a fixed point**. A system initially at a fixed point stays there. This is the same definition as in Chapter 2 for first order systems, i.e. at a fixed point the velocity field is zero.

Since $(x_1, x_2) \to (1,2)$ as $t \to \infty$, using the language of Chapter 2 this is called a **stable fixed point**. In the analysis of first order systems, the determination of fixed points and the behaviour near them gives a great deal of information about the motion of the system. The same ideas carry over to second order systems. We investigate such motions by drawing a graph of x_2 against x_1.

Equations (3.4) are parametric equations for the dependent variables x_1 and x_2. Figure 3.2 shows a graph of x_2 against x_1.

A solution curve along which the coordinates x_1 and x_2 vary as t increases is called a **trajectory, phase path** or **orbit**. The plane containing the solution curves is called **the phase plane**. The phase point initially at the point (0,3) of the phase plane moves along the trajectory approaching the fixed point (1,2) in an infinite time.

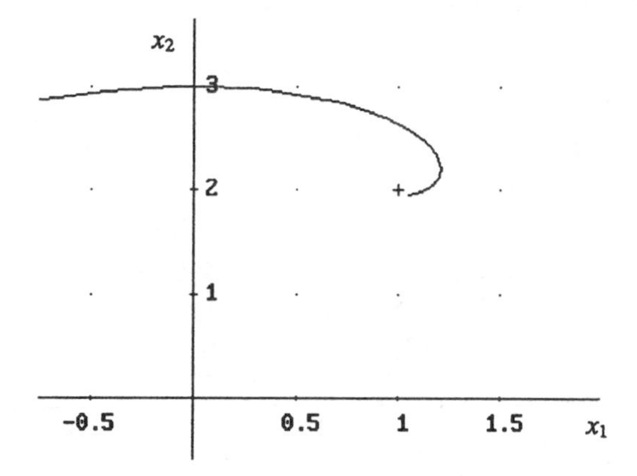

Fig 3.2 A trajectory, phase path or orbit

The trajectory in Fig 3.2 was drawn from an analytical solution of the system of differential equations. For non-linear systems such analytical solutions cannot normally be found. It is sometimes possible to obtain the equation of the trajectory by converting the pair of first order differential equations into a single first order differential equation. For example, consider the system

$$\frac{dx_1}{dt} = x_2$$

$$\frac{dx_2}{dt} = x_1$$

Dividing the second equation by the first gives

$$\frac{dx_2/dt}{dx_1/dt} = \frac{dx_2}{dt}\frac{dt}{dx_1} = \frac{dx_2}{dx_1} = \frac{x_1}{x_2}$$

The first order differential equation

$$\frac{dx_2}{dx_1} = \frac{x_1}{x_2}$$

is variables separable and can be solved easily:

$$\int x_2 dx_2 = \int x_1 dx_1$$

$$\frac{x_2^2}{2} = \frac{x_1^2}{2} + c$$

$$x_2 = \pm\sqrt{x_1^2 + 2c}$$

where c is a constant. Figure 3.3 shows several trajectories for this system, the orbits are hyperbolas.

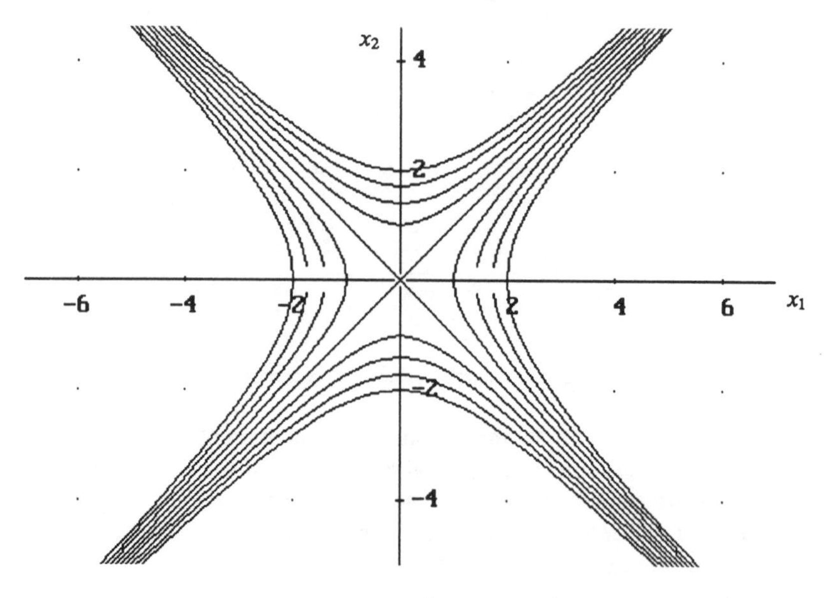

Fig 3.3 Trajectories for $\dfrac{dx_1}{dt} = x_2$ and $\dfrac{dx_2}{dt} = x_1$

More generally, consider the autonomous system

$$\frac{dx_1}{dt} = v_1(x_1, x_2)$$
$$\frac{dx_2}{dt} = v_2(x_1, x_2) \tag{3.5}$$

Dividing the second equation by the first equation we obtain

$$\frac{dx_2}{dx_1} = \frac{v_2(x_1, x_2)}{v_1(x_1, x_2)} \tag{3.6}$$

(Normally this equation will only be meaningful if the system is autonomous because the variable t has been explicitly eliminated.)

The trajectories for the system are solutions of differential equation (3.5). To construct the phase diagram for system (3.5) we would draw at a chosen number of points (x_1,x_2) the vector $v_1\mathbf{i} + v_2\mathbf{j}$ which has the same slope as the solution of equation (3.6). The set of vectors forming the phase diagram is the same as the set of slopes obtained from the direction field of equation (3.6). Thus the direction field will provide a geometric view of the trajectories for an autonomous system. The direction of motion of a phase point can be obtained by adding arrows to the direction field. Example 2 shows the method of approach.

Example 2

Formulate the equation of the trajectories for the system

$$\frac{dx_1}{dt} = -x_1 + x_2 - 1$$
$$\frac{dx_2}{dt} = -x_1 - x_2 + 3 \qquad (3.7)$$

Draw the associated direction field and sketch in some solution curves near the fixed point (1,2).

Solution

The differential equation for the trajectories is

$$\frac{dx_2}{dx_1} = \frac{-x_1 - x_2 + 3}{-x_1 + x_2 - 1} \qquad (3.8)$$

Figure 3.4 shows the direction field and some solution curves for this differential equation.

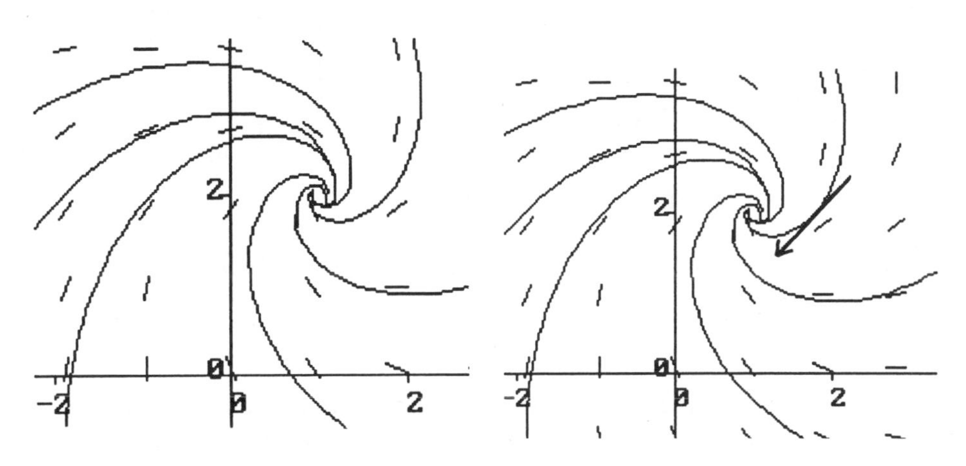

Fig 3.4 Direction field for equation (3.8)

Fig 3.5 Putting directions on the solution curve

To add directions to the solution curves we return to the original set of equations (3.7). Consider the point (2,2) on the solution curve AB and substitute $x_1 = 2$, $x_2 = 2$ in equations (3.7). We have at this point $\frac{dx_1}{dt} = -1$ and $\frac{dx_2}{dt} = -1$. The velocity field is in the direction of the vector $-\mathbf{i}-\mathbf{j}$. Figure 3.5 shows the solution curve through the point (2,2) and the vector $-\mathbf{i}-\mathbf{j}$. From this vector we can infer the direction of motion for a phase point on the trajectory AB.

In a similar way, directions can be included on all the solution curves. So now we can give a qualitative description of the motion of the system. Any phase point initially near the fixed point will spiral into the fixed point in an infinite time.

TUTORIAL PROBLEM 3.3

Figure 3.6 shows the associated direction field and some solution curves for the systems

(i) $$\frac{dx_1}{dt} = 2x_1x_2 \qquad \frac{dx_2}{dt} = x_2^2 - x_1^2$$

(ii) $$\frac{dx_1}{dt} = 3x_1 + x_2 \qquad \frac{dx_2}{dt} = 2x_1 + 4x_2$$

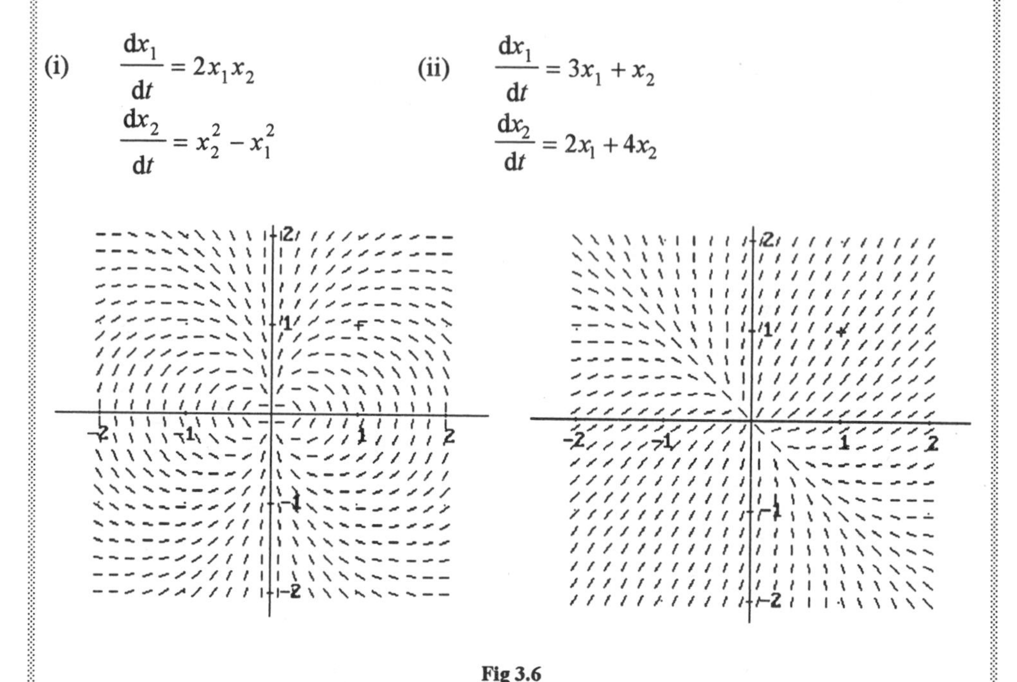

Fig 3.6

Add directions to the solution curves and hence give a qualitative description for the motion of each system.

In this tutorial problem each system has a fixed point at the origin. For system (i) the fixed point appears stable for phase paths in the region $x_2 < 0$ and unstable for $x_2 > 0$. This fixed point appears similar to a shunt in one-dimensional systems in that motion on the phase paths enters on one side and leaves from the other. Compare this with system (ii) in which motion on the phase curves all leave the fixed point. This is an example of an unstable fixed point.

The first step in analysing a dynamical system is to find the positions of the fixed points, by solving the equation $\mathbf{v} = \mathbf{0}$.

Example 3

Find the fixed points of the non-linear second order system

$$\frac{dx_1}{dt} = 2x_1x_2 + x_1$$

$$\frac{dx_2}{dt} = x_2^2 - x_1^2$$

Solution

The fixed points are given by solving the equations

$$2x_1x_2 + x_1 = 0 \tag{3.9}$$

$$x_2^2 - x_1^2 = 0 \tag{3.10}$$

From equation (3.9) either $x_1 = 0$ or $x_2 = -0.5$. Substituting these values into equation (3.10) we have

$$x_1 = 0 \quad \Rightarrow \quad x_2 = 0$$

$$x_2 = -0.5 \quad \Rightarrow \quad x_1 = 0.5 \text{ or } -0.5$$

There are three fixed points for this system (0,0), (0.5,–0.5) and (–0.5,–0.5).

TUTORIAL PROBLEM 3.4

Find all the fixed points of the system

$$\frac{dx_1}{dt} = 2 - x_1x_2$$

$$\frac{dx_2}{dt} = x_2(1 - x_1) - 3$$

Having found the positions of the fixed points of a system the next step is to classify them. In Chapter 2 the fixed points of a first order system were classified as stable, unstable or a shunt. For second order systems similar classifications are possible, however there are more interesting types of fixed point. The next section investigates the full range of possible fixed points; here we will just introduce three simple types of fixed point.

A fixed point $\mathbf{x}_f$ is called an **attractor** of some motion if

$$\lim_{t \to +\infty} \mathbf{x}(t) = \mathbf{x}_f$$

A fixed point $\mathbf{x}_f$ is called a **repellor** of some motion if

$$\lim_{t \to -\infty} \mathbf{x}(t) = \mathbf{x}_f$$

Figure 3.7 shows the trajectory for an attractor and repellor.

Fig 3.7 A trajectory which is an attractor (a) and a repellor (b)

For some motions the fixed point is an attractor for **all** motions passing sufficiently close to it. What this implies for the motion of the system is that if initially a phase point is close to the fixed point then its motion will move towards the fixed point. In this case the fixed point is said to be **strongly** or **asymptotically stable**. A fixed point which is a repellor for **all** motions passing sufficiently close to it is said to be **strongly** or **asymptotically unstable**.

It is possible for a fixed point to be both an attractor for some motions and a repellor for others. Figure 3.8 shows such a situation; this is the system (a) given in Tutorial Problem 3.3. Such a fixed point is semistable. It is the equivalent of a shunt for a first order system.

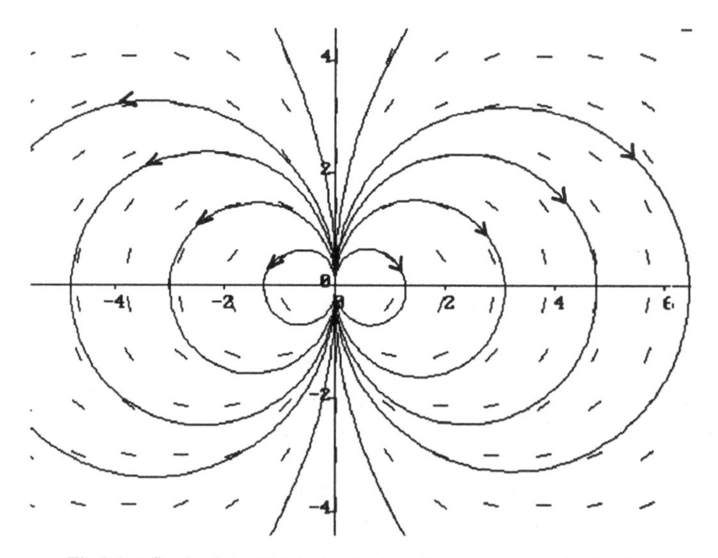

Fig 3.8 A fixed point which is simultaneously an attractor and a repellor

TUTORIAL PROBLEM 3.5

Figure 3.9 shows some trajectories and four fixed points, A, B, C and D, of a second order autonomous system.

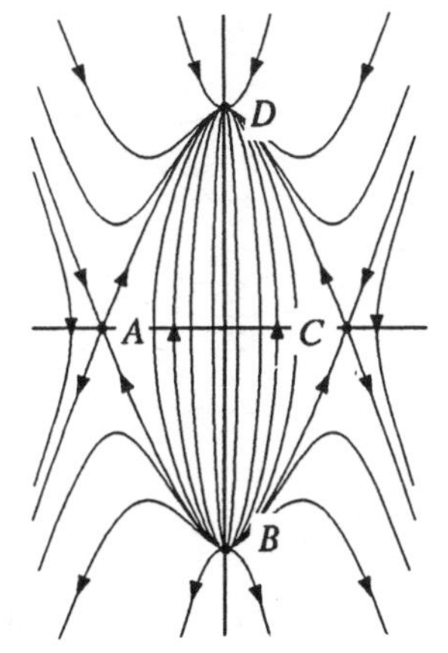

Fig 3.9 Trajectories near four fixed points

(i) Identify the fixed points which attract **some** motion.

(ii) Identify the fixed points which repel **some** motion.

(iii) Classify the fixed points.

There is a third type of simple fixed point called a **centre**. Consider the system

$$\frac{dx_1}{dt} = -\alpha x_2$$

$$\frac{dx_2}{dt} = \beta x_1$$

with fixed point (0,0) where α and β are constants. The equation for the trajectories is given by

$$\frac{dx_2}{dx_1} = \frac{-\beta x_1}{\alpha x_2}$$

which can be solved analytically to give

$$\frac{x_1^2}{\alpha} + \frac{x_2^2}{\beta} = c$$

where c is a constant. This is the equation of a system of ellipses centred on the fixed point. Figure 3.10 shows some of the trajectories. If $\alpha = \beta$ then the ellipses become circles.

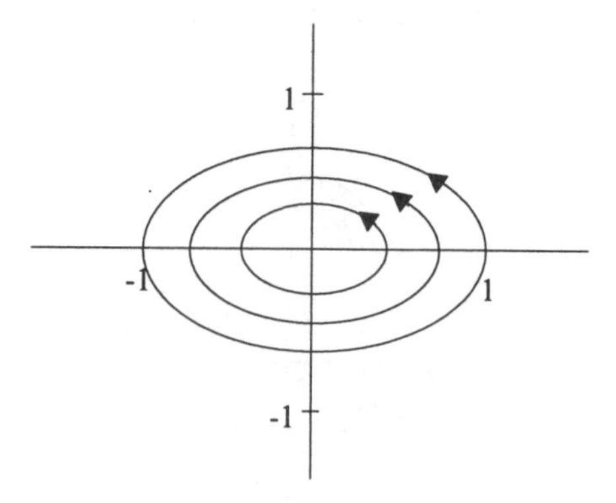

Fig 3.10 An example of a centre

This fixed point is stable because any phase point sufficiently close to the centre stays in orbit close to the fixed point. A centre is not strongly stable because it is not an attractor for any motion.

TUTORIAL PROBLEM 3.6

A system is modelled by the second order differential equation

$$\frac{d^2x}{dt^2} + x = 0$$

(i) Show that defining $y = \dfrac{dx}{dt}$ reduces it to the pair of simultaneous first order differential equations

$$\frac{dx}{dt} = y, \qquad \frac{dy}{dt} = -x$$

(ii) Verify that the solution of these equations satisfying $x = x_0$, $y = y_0$ when $t = 0$ is

$$x(t) = x_0 \cos t + y_0 \sin t, \quad y(t) = -x_0 \sin t + y_0 \cos t$$

(iii) Show that by introducing polar coordinates in the phase space, $x = r\cos\theta$, $y = r\sin\theta$, the equations of motion become

$$\dot{r} = 0, \qquad \dot{\theta} = -1$$

(iv) Sketch the solution curves. Give a qualitative description of the motion of the system.

Exercises 3.3

1. For the following phase planes identify the fixed points which (a) attract **some** motion; (b) repel **some** motion. Classify the fixed points.

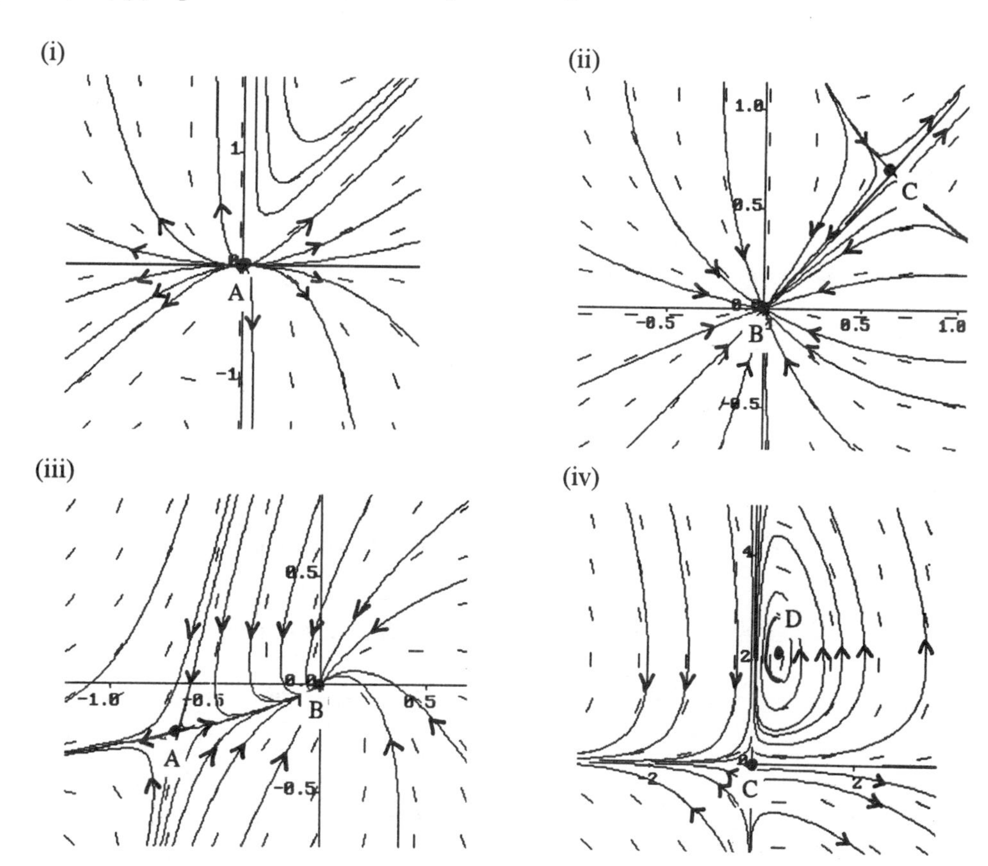

2. For each of the following second order systems find the fixed points. Use an appropriate computer software package to draw the phase diagrams and trajectories for each system. Describe qualitatively the motion of each system for a phase initially close to the fixed point.

(i) $\dfrac{dx_1}{dt} = 3x_1 + 4x_2,\qquad \dfrac{dx_2}{dt} = 2x_1 + x_2$

(ii) $\dfrac{dx_1}{dt} = x_1 + 2x_2,\qquad \dfrac{dx_2}{dt} = -2x_1 + 5x_2 - 9$

(iii) $\dfrac{dx_1}{dt} = x_1^2 + x_2^2 + 2x_1,\qquad \dfrac{dx_2}{dt} = x_1 x_2 + x_2$

(iv) $$\frac{dx_1}{dt} = 2x_1x_2 - 4x_2 - 8, \qquad \frac{dx_2}{dt} = -x_1^2 + 4x_2^2$$

(v) $$\frac{dx_1}{dt} = -x_1^2 + x_2 + 2, \qquad \frac{dx_2}{dt} = 2\left(x_1^2 - x_2^2\right)$$

3. Show that the second order autonomous system obtained from the first order non-autonomous system

$$\frac{dx}{dt} = v(x,t)$$

has no fixed points.

4. Consider the system with equations of motion

$$\frac{dx}{dt} = y, \qquad \frac{dy}{dt} = x,$$

corresponding to the second order differential equation $\frac{d^2x}{dt^2} - x = 0$.

(i) Show that the origin is a fixed point of this system.

(ii) Verify that the solution of the system which satisfies the initial conditions $(x(0), y(0)) = (x_0, y_0)$ is

$$x(t) = \frac{1}{2}(x_0 + y_0)e^t + \frac{1}{2}(x_0 - y_0)e^{-t}$$

$$y(t) = \frac{1}{2}(x_0 + y_0)e^t - \frac{1}{2}(x_0 - y_0)e^{-t}$$

(iii) Deduce that the origin is an attractor for the motion of any point on the line $y = -x$, but a repellor for the motion of any other point. Conclude that the origin is an unstable fixed point.

5. A second order system is modelled by the differential equation

$$\frac{d^2x}{dt^2} + \left(\frac{dx}{dt}\right)^2 + x = 1$$

Convert this equation to a pair of simultaneous first order equations.

Find the fixed points of this system. Sketch the direction field and some solution curves. Hence give a qualitative description of the motion of the system.

6. A second order system is modelled by the differential equation

$$\frac{d^2x}{dt^2}+\frac{dx}{dt}+x-x^3=0$$

Convert this equation to a pair of simultaneous first order equations using the two transformations

(i) $x_1=x$ and $x_2=\dfrac{dx}{dt}$, and

(ii) $x_1=x$ and $x_2=\dfrac{dx}{dt}+x$.

For each new pair of simultaneous first order differential equations find the fixed points of this system. Sketch the direction field and some solution curves. Hence give a qualitative description of the motion of the system.

Are your two descriptions the same in each case?

3.4 Classification of fixed points of linear systems

A dynamical system is **linear** if it is modelled by a system of first order linear constant coefficient homogeneous differential equations. For a second order system the general form of such a system has the standard form

$$\begin{aligned}\frac{dx_1}{dt}&=ax_1+bx_2\\ \frac{dx_2}{dt}&=cx_1+dx_2\end{aligned} \tag{3.11}$$

which can be written in matrix/vector form as

$$\frac{d\mathbf{x}}{dt}=\mathbf{Ax} \tag{3.12}$$

where **A** is the matrix of coefficients $\mathbf{A}=\begin{bmatrix}a & b\\ c & d\end{bmatrix}$.

As we shall see in the next section, these linear systems play a key role in describing the nature and solution of non-linear systems.

A fixed point of the system (3.12) is a solution of the equation

$$\mathbf{Ax} = \mathbf{0}$$

If the matrix **A** is non-singular (i.e. **A** has an inverse) then the origin is the only solution i.e.

$$\mathbf{x}_f = \mathbf{0}$$

If the matrix **A** is singular then there is a non-zero $\mathbf{x}_f$ such that $\mathbf{Ax}_f = \mathbf{0}$. In fact any scalar multiple of $\mathbf{x}_f$, i.e. $\mathbf{x} = \alpha\mathbf{x}_f$, is also a solution of $\mathbf{Ax} = \mathbf{0}$. Thus there are a whole line of fixed points passing through the origin.

A fixed point is said to be **isolated** if there are no other fixed points in its neighbourhood. A fixed point is isolated if **A** is non-singular, i.e. if $\mathbf{A}^{-1}$ exists. A linear system with non-singular matrix of coefficients is said to be **simple**. A simple linear system has an isolated fixed point at the origin.

TUTORIAL PROBLEM 3.7

Consider the system

$$\frac{d\mathbf{x}}{dt} = \mathbf{Ax} + \mathbf{h}$$

where $\mathbf{h} = \begin{bmatrix} e \\ f \end{bmatrix}$.

(i) Show that the fixed point is at $\mathbf{x}_f = -\mathbf{A}^{-1}\mathbf{h}$.

(ii) Show that the transformation $\mathbf{X} = \mathbf{x} - \mathbf{x}_f$ reduces the system to the standard homogeneous form.
If necessary describe the geometrical significance of this transformation.

This tutorial problem shows that any linear system with an isolated fixed point can be transformed to one with the same matrix of coefficients and whose fixed point is at the origin. We shall see that the classification and behaviour of the system (3.12) depends on the eigenvalues and eigenvectors of the matrix **A**. The steps in the analysis are now explored through three important properties.

The implications of the solution of this problem are important; it means that we can always reduce the system 3.12 to a homogeneous system

$$\frac{d\mathbf{x}}{dt} = \mathbf{Ax}$$

Property 3.1

The general solution of the simple linear system

$$\frac{d\mathbf{x}}{dt} = \mathbf{Ax}$$

is given by

$$\mathbf{x} = \mathbf{u}_1 e^{\lambda_1 t} + \mathbf{u}_2 e^{\lambda_2 t}$$

where λ_1 and λ_2 are the eigenvalues of $\mathbf{A}$, and $\mathbf{u}_1$ and $\mathbf{u}_2$ are the corresponding eigenvectors.

Let $\mathbf{x} = \mathbf{u}e^{\lambda t}$ where λ is a constant scalar and $\mathbf{u}$ is a constant vector; then $\frac{d\mathbf{x}}{dt} = \lambda \mathbf{u} e^{\lambda t}$.

Substituting into $\frac{d\mathbf{x}}{dt} = \mathbf{Ax}$ gives

$$\lambda \mathbf{u} e^{\lambda t} = \mathbf{Au} e^{\lambda t}$$

Hence λ and $\mathbf{u}$ satisfy the equation

$$\mathbf{Au} = \lambda \mathbf{u}$$

This equation defines the eigenvalues λ and corresponding eigenvectors $\mathbf{u}$ of the matrix $\mathbf{A}$. In general a second order matrix will have two eigenvalues and corresponding eigenvectors so the general solution will be

$$\mathbf{x} = \mathbf{u}_1 e^{\lambda_1 t} + \mathbf{u}_2 e^{\lambda_2 t}$$

Property 3.2

Let $\mathbf{u}_1$ and $\mathbf{u}_2$ be the eigenvectors corresponding to eigenvalues λ_1 and λ_2 of the matrix $\mathbf{A}$ for the simple linear system

$$\frac{d\mathbf{x}}{dt} = \mathbf{Ax}$$

If $\mathbf{P} = [\mathbf{u}_1 \ \ \mathbf{u}_2]$ is the matrix whose columns are the eigenvectors $\mathbf{u}_1$ and $\mathbf{u}_2$ then $\mathbf{P}^{-1}\mathbf{AP} = \mathbf{D}$, where $\mathbf{D}$ is the diagonal matrix

$$\mathbf{D} = \begin{bmatrix} \lambda_1 & 0 \\ 0 & \lambda_2 \end{bmatrix}$$

To show this result we begin with the product **AP**,

$$\mathbf{AP} = \mathbf{A}[\mathbf{u}_1\ \ \mathbf{u}_2] = [\mathbf{A}\mathbf{u}_1\ \ \mathbf{A}\mathbf{u}_2] = [\lambda_1\mathbf{u}_1\ \ \lambda_2\mathbf{u}_2]$$

since $\mathbf{Au} = \lambda\mathbf{u}$ defines the eigenvalues and eigenvectors. Now we can write

$$[\lambda_1\mathbf{u}_1\ \ \lambda_2\mathbf{u}_2] = [\mathbf{u}_1\ \ \mathbf{u}_2]\begin{bmatrix} \lambda_1 & 0 \\ 0 & \lambda_2 \end{bmatrix} = \mathbf{PD}$$

Hence

$$\mathbf{AP} = \mathbf{PD}$$

Multiplying by $\mathbf{P}^{-1}$ on the left of each side gives the required result

$$\mathbf{P}^{-1}\mathbf{AP} = \mathbf{D}$$

In general, the second order system $\dfrac{d\mathbf{x}}{dt} = \mathbf{Ax}$ is a pair of coupled simultaneous differential equations with each equation containing both variables x_1 and x_2. The matrix of eigenvectors **P** allows us to transform the system to a decoupled system.

Let $\mathbf{x} = \mathbf{Py}$; then $\dfrac{d\mathbf{x}}{dt} = \mathbf{P}\dfrac{d\mathbf{y}}{dt}$ and

$$\frac{d\mathbf{x}}{dt} = \mathbf{P}\frac{d\mathbf{y}}{dt} = \mathbf{APy}$$

Multiplying by $\mathbf{P}^{-1}$ gives

$$\frac{d\mathbf{y}}{dt} = \mathbf{P}^{-1}\mathbf{APy} = \mathbf{Dy}$$

The two equations for **y** are

$$\frac{dy_1}{dt} = \lambda_1 y_1 \quad \text{and} \quad \frac{dy_2}{dt} = \lambda_2 y_2$$

The decoupled system $\dfrac{d\mathbf{y}}{dt} = \mathbf{Dy}$, called a **canonical system**, is much easier to solve and analyse than the original system.

TUTORIAL PROBLEM 3.8

Find the eigenvalues and corresponding eigenvectors of the following matrices. In each case, write down the matrix **P** and show that $\mathbf{P}^{-1}\mathbf{AP}$ is diagonal.

(i) $\mathbf{A} = \begin{bmatrix} 3 & 1 \\ 2 & 4 \end{bmatrix}$ (ii) $\mathbf{A} = \begin{bmatrix} 0 & -1 \\ 4 & 4 \end{bmatrix}$ (iii) $\mathbf{A} = \begin{bmatrix} 2 & 1 \\ -2 & 4 \end{bmatrix}$

This tutorial problem introduces the three cases that can occur: (i) two real eigenvalues, (ii) equal eigenvalues, and (iii) complex eigenvalues. In the case of complex eigenvalues we will find that the diagonal matrix **D** is not the best canonical system to work with. We consider the three cases separately.

(a) Two real eigenvalues

The canonical form is $\dfrac{d\mathbf{y}}{dt} = \mathbf{Dy}$ where $\mathbf{D} = \begin{bmatrix} \lambda_1 & 0 \\ 0 & \lambda_2 \end{bmatrix}$. The solutions for y_1 and y_2 are

$$y_1 = C_1 e^{\lambda_1 t} \quad \text{and} \quad y_2 = C_2 e^{\lambda_2 t}$$

where c_1 and c_2 are constants. The phase curves can be obtained by eliminating t from these equations to give

$$y_2 = c_2 \left(\frac{y_1}{c_1} \right)^{\lambda_2/\lambda_1}$$

The behaviour of the system depends on the signs and relative sizes of λ_1 and λ_2. Figure 3.11 shows the basic types and their names.

For the saddle points the coordinate axes (excluding the fixed point at the origin) are solutions called **separatrices**. All the phase curves have these separatrices as asymptotes.

(b) Equal eigenvalues

If there are two linearly independent eigenvectors then

$$\frac{d\mathbf{y}}{dt} = \mathbf{Dy} \quad \text{where} \quad \mathbf{D} = \begin{bmatrix} \lambda & 0 \\ 0 & \lambda \end{bmatrix}$$

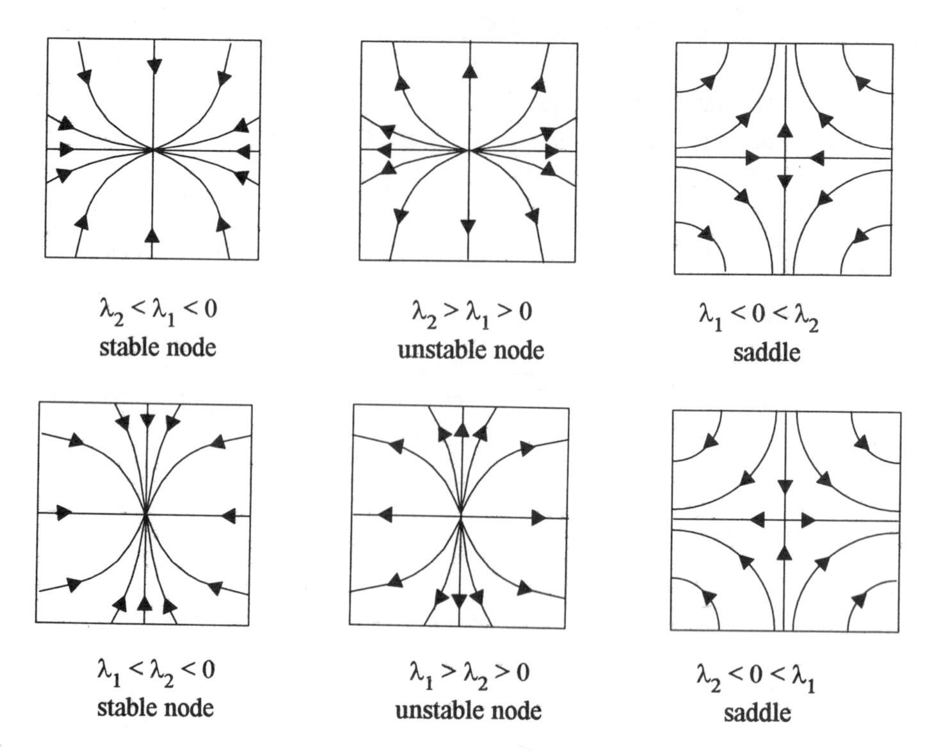

Fig 3.11 Phase curves near a fixed point for real distinct eigenvalues

The equations of motion have solutions

$$y_1 = c_1 e^{\lambda t} \quad \text{and} \quad y_2 = c_2 e^{\lambda t}$$

giving the equations of the phase curves as the radial straight lines

$$\frac{y_2}{y_1} = \frac{c_2}{c_1}$$

Figure 3.12 shows the phase curves in this case.

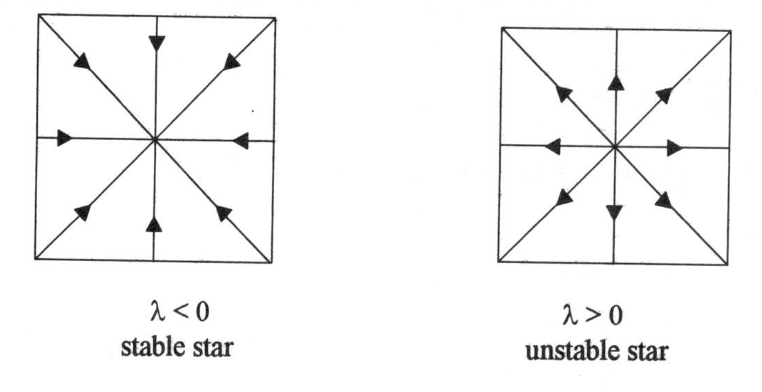

Fig 3.12 Phase curves near a fixed point for equal eigenvalues and two independent eigenvectors

If there are not two linearly independent eigenvectors then the matrix $\mathbf{P} = [\mathbf{u}\ \ \mathbf{v}]$, where $\mathbf{v}$ is any vector so that $\mathbf{P}$ is non-singular, will transform the matrix $\mathbf{A}$ to the canonical form $\mathbf{M} = \begin{bmatrix} \lambda & 1 \\ 0 & \lambda \end{bmatrix}$.

In this case the equations of motion are

$$\frac{\mathrm{d}y_1}{\mathrm{d}t} = \lambda y_1 + y_2$$

$$\frac{\mathrm{d}y_2}{\mathrm{d}t} = \lambda y_2$$

The solutions for y_1 and y_2 are

$$y_2 = c_2 \mathrm{e}^{\lambda t} \quad \text{and} \quad y_1 = (c_1 + c_2 t)\mathrm{e}^{\lambda t}$$

The equations of the phase curves are given by

$$y_1 = y_2\left(\frac{c_1}{c_2} + \frac{1}{\lambda}\ln\left(\frac{y_2}{c_2}\right)\right)$$

Figure 3.13 shows the phase curves in this case.

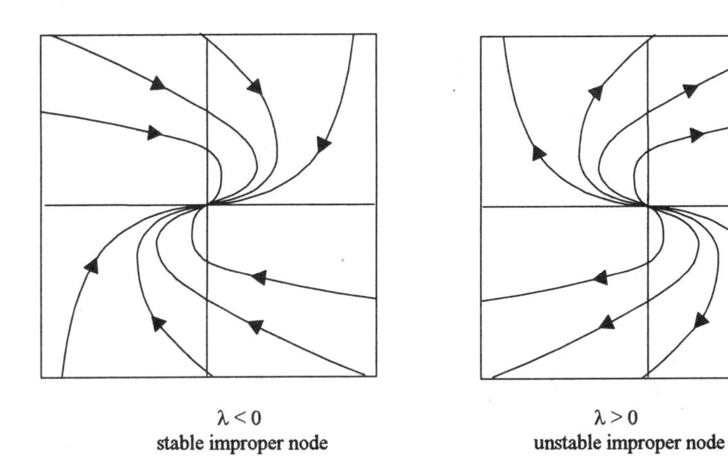

Fig 3.13 Phase curves near a fixed point for equal eigenvalues

(c) Complex eigenvalues

When the eigenvalues are complex the diagonal/canonical form contains complex numbers and an alternative approach is sought.

The canonical form in this case is the matrix $\mathbf{M} = \begin{bmatrix} \mu & -\omega \\ \omega & \mu \end{bmatrix}$ where μ and $\pm\omega$ are real and imaginary parts of the (complex) eigenvalues $\lambda = \mu \pm i\omega$.

TUTORIAL PROBLEM 3.9

(i) Show that if the matrix **A** has complex eigenvalues then $\mu = \frac{1}{2}(a+d)$ and $\omega = \frac{1}{2}\sqrt{4(ad-bc)-(a+d)^2}$.

(ii) Show that if $\mathbf{P} = \begin{bmatrix} a-\mu & -\omega \\ c & 0 \end{bmatrix}$ then $\mathbf{P}^{-1}\mathbf{AP} = \begin{bmatrix} \mu & -\omega \\ \omega & \mu \end{bmatrix}$.

The canonical form for this case involves the pair of differential equations

$$\frac{\mathrm{d}y_1}{\mathrm{d}t} = \mu y_1 - \omega y_2$$

$$\frac{\mathrm{d}y_2}{\mathrm{d}t} = \omega y_1 + \mu y_2$$

To solve this kind of system we introduce polar coordinates

$$y_1 = r\cos\theta \quad \text{and} \quad y_2 = r\sin\theta$$

giving

$$\frac{\mathrm{d}r}{\mathrm{d}t}\cos\theta - r\sin\theta\frac{\mathrm{d}\theta}{\mathrm{d}t} = \mu r\cos\theta - \omega r\sin\theta \tag{3.13}$$

$$\frac{\mathrm{d}r}{\mathrm{d}t}\sin\theta + r\cos\theta\frac{\mathrm{d}\theta}{\mathrm{d}t} = \omega r\cos\theta + \mu r\sin\theta \tag{3.14}$$

$$(3.13) \times \cos\theta + (3.14) \times \sin\theta \quad \Rightarrow \quad \frac{\mathrm{d}r}{\mathrm{d}t} = \mu r \quad \Rightarrow \quad r = r_0 e^{\mu t}$$

$$(3.14) \times \cos\theta - (3.13) \times \sin\theta \quad \Rightarrow \quad r\frac{\mathrm{d}\theta}{\mathrm{d}t} = \omega r \quad \Rightarrow \quad \theta = \theta_0 + \omega t$$

Eliminating t between these equations for r and θ gives

$$r = r_0 e^{\mu(\theta_0 + \theta)/\omega}$$

which are spirals if $\mu \neq 0$ and circles of $\mu = 0$. The phase curves are shown in Fig 3.14.

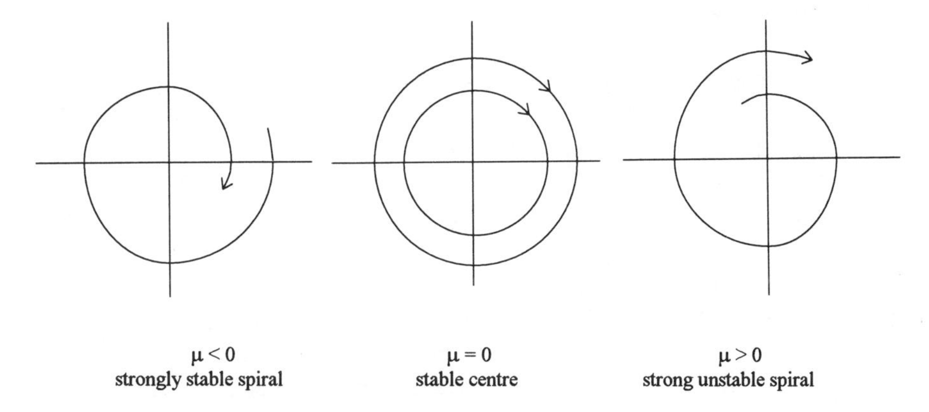

$\mu < 0$	$\mu = 0$	$\mu > 0$
strongly stable spiral	stable centre	strong unstable spiral

Fig 3.14 Phase curves near a fixed point for complex eigenvalues

We conclude that the classification of the fixed points and the behaviour of the phase curves near the fixed points for a simple linear system depend on the form of the eigenvalues.

Property 3.3

The eigenvalues of a matrix **A** are real distinct, equal or complex depending on the sign of $(\mathrm{tr}(\mathbf{A}))^2 - 4\det(\mathbf{A})$, where $\mathrm{tr}(\mathbf{A}) = (a + d)$ and $\det(\mathbf{A}) = (ad - bc)$.

To show this result suppose that

$$\mathbf{A} = \begin{bmatrix} a & b \\ c & d \end{bmatrix}$$

Then the eigenvalues of **A** are given by

$$\begin{vmatrix} a - \lambda & b \\ c & d - \lambda \end{vmatrix} = 0$$

$$\Rightarrow \quad (a - \lambda)(d - \lambda) - bc = 0$$

$$\Rightarrow \quad \lambda^2 - (a + d)\lambda + ad - bc = 0$$

Solving for λ we have

$$\lambda = (a+d) \pm \sqrt{(a+d)^2 - 4(ad-bc)}$$
$$\lambda = \mathrm{tr}(\mathbf{A}) \pm \sqrt{(\mathrm{tr}(\mathbf{A})^2 - 4\det(\mathbf{A})}$$

The eigenvalues are

(a) real distinct if $(\mathrm{tr}(\mathbf{A}))^2 - 4\det(\mathbf{A}) > 0$

(b) equal if $(\mathrm{tr}(\mathbf{A}))^2 - 4\det(\mathbf{A}) = 0$

(c) complex if $(\mathrm{tr}(\mathbf{A}))^2 - 4\det(\mathbf{A}) < 0$

Hence the type of fixed point depends on the value of this discriminant. Figure 3.15 contains a summary of the 10 basic types of fixed point drawn on the det(**A**) vs. tr(**A**) plane. It shows how we can move straight from the values of the trace and determinant of **A** to a description of the phase curves near a simple fixed point.

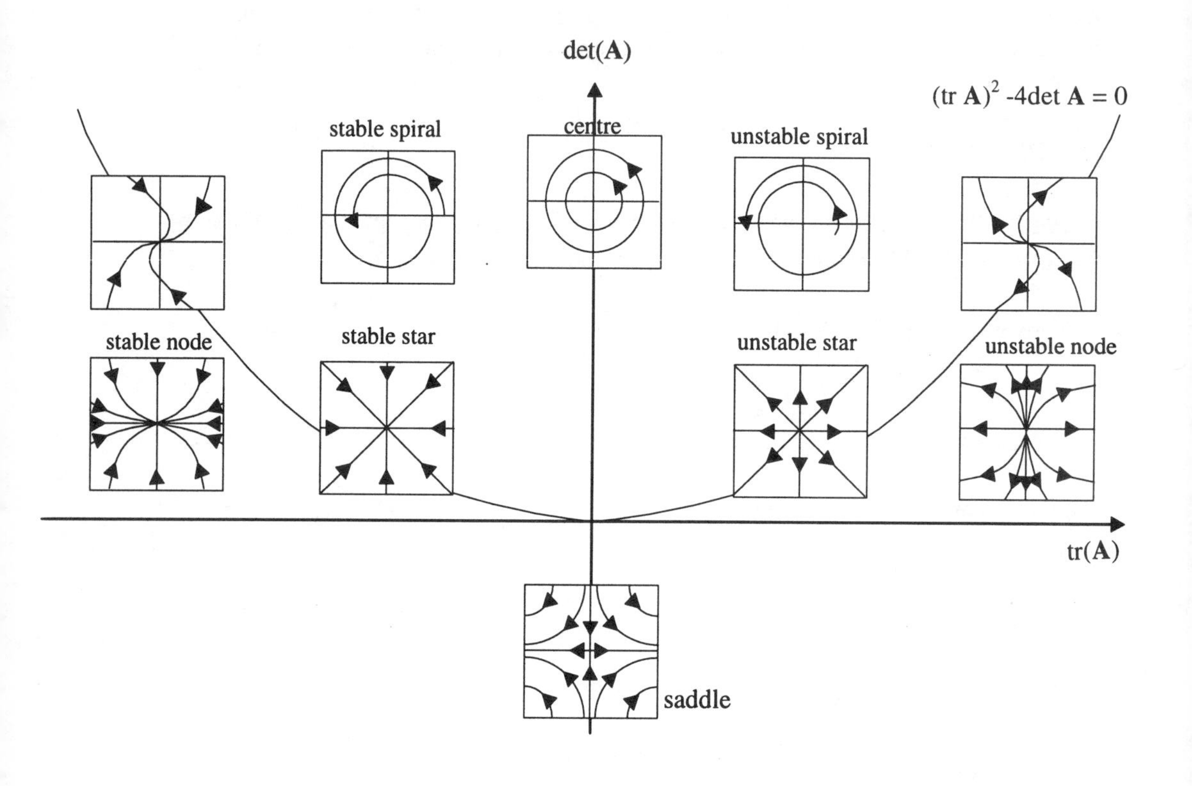

Fig 3.15 Summary of the types of fixed point in the det(**A**) – tr(**A**) plane

Example 4

Consider the second order system

$$\frac{dx_1}{dt} = x_1 + x_2 + 1 \qquad \frac{dx_2}{dt} = 2x_1 - x_2 + 5$$

(i) Find the position of the fixed point of the system.

(ii) Classify the equivalent canonical system.

(iii) Find the eigenvalues and eigenvectors of the system and use the analytical solution of the canonical system to find the general solution for $\mathbf{x}$.

Solution

(i) The fixed point of the system is given by

$$\begin{aligned} x_1 + x_2 + 1 &= 0 \\ 2x_1 - x_2 + 5 &= 0 \end{aligned}$$

Solving for x_1 and x_2 gives $x_1 = -2$ and $x_2 = 1$. There is a fixed point at $(-2,1)$.

(ii) The matrix of coefficients is $\mathbf{A} = \begin{bmatrix} 1 & 1 \\ 2 & -1 \end{bmatrix}$. For this matrix $\mathrm{tr}(\mathbf{A}) = 0$ and $\det(\mathbf{A}) = -3$. The equivalent canonical system has a saddle point at the origin (see Fig 3.15).

(iii) The eigenvalues of $\mathbf{A}$ are given by

$$\begin{vmatrix} 1-\lambda & 1 \\ 2 & -1-\lambda \end{vmatrix} = 0 \quad \Rightarrow \quad (1-\lambda)(-1-\lambda) - 2 = 0$$

$$\Rightarrow \quad \lambda^2 - 3 = 0$$

Solving for λ we have $\lambda = \pm\sqrt{3}$. The canonical system has general solution

$$y_1 = c_1 e^{\sqrt{3}t} \quad \text{and} \quad y_2 = c_2 e^{-\sqrt{3}t}$$

The eigenvectors are solutions of $\mathbf{Au} = \lambda\mathbf{u}$.

For $\lambda = \sqrt{3}$, $\begin{bmatrix} 1-\sqrt{3} & 1 \\ 2 & -1-\sqrt{3} \end{bmatrix} \begin{bmatrix} u_1 \\ u_2 \end{bmatrix} = 0,$

so the eigenvector is $[1, \sqrt{3}-1]^{\mathrm{T}}$.

For $\lambda = -\sqrt{3}$, $\begin{bmatrix} 1+\sqrt{3} & 1 \\ 2 & -1+\sqrt{3} \end{bmatrix} \begin{bmatrix} u_1 \\ u_2 \end{bmatrix} = 0$,

so the eigenvector is $[1, -\sqrt{3}-1]^T$.

The transformation matrix **P** is $\mathbf{P} = \begin{bmatrix} 1 & 1 \\ \sqrt{3}-1 & -\sqrt{3}-1 \end{bmatrix}$.

The general solution for **x** is $\mathbf{x}_f + \mathbf{P}\mathbf{y}$, so

$$x_1 = c_1 e^{\sqrt{3}t} + c_2 e^{-\sqrt{3}t} - 2$$

$$x_2 = (\sqrt{3}-1)c_1 e^{\sqrt{3}t} - (\sqrt{3}+1)c_2 e^{-\sqrt{3}t} + 1$$

Figure 3.16 shows some of the phase curves near the origin in the y_2–y_1 plane and near the fixed point (–2,1) in the x_2–x_1 plane.

The fixed point is translated to coincide with the origin by transforming the system to $\frac{d\mathbf{x}}{dt} = \mathbf{A}\mathbf{x}$, i.e. ignoring the vector $\begin{bmatrix} 1 \\ 5 \end{bmatrix}$.

The matrix **P** formed from the eigenvectors of **A** is equivalent to a mapping that orientates the separatrices of the saddle point along the direction of the eigenvectors of **A**. Although the phase portrait is distorted, the fixed point is still a saddle and the qualitative behaviour near the fixed point remains the same.

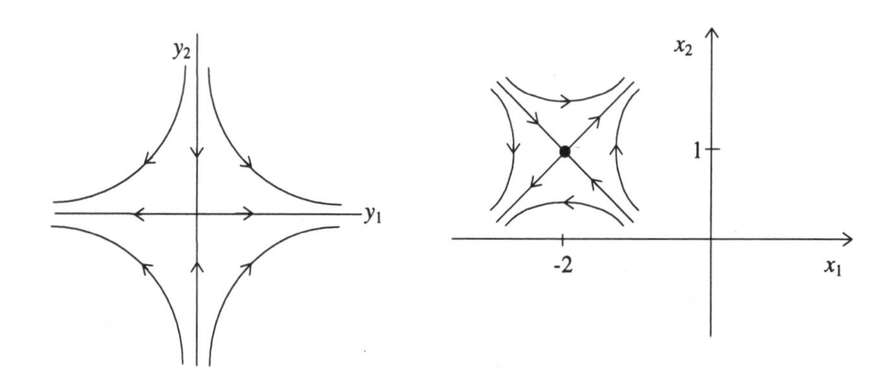

Fig 3.16 The matrix **P** produces a rotation of the coordinate axes

Exercises 3.4

1. Sketch the phase portraits for the linear system

$$\frac{d\mathbf{x}}{dt} = \mathbf{A}\mathbf{x}$$

where **A** is given by the following matrices:

(i) $\begin{bmatrix} -1 & 0 \\ 0 & -2 \end{bmatrix}$ (ii) $\begin{bmatrix} 1 & 1 \\ 0 & 2 \end{bmatrix}$ (iii) $\begin{bmatrix} 3 & 0 \\ 0 & 1 \end{bmatrix}$

(iv) $\begin{bmatrix} -2 & 0 \\ 0 & 3 \end{bmatrix}$ (v) $\begin{bmatrix} -1 & 2 \\ -2 & -1 \end{bmatrix}$ (vi) $\begin{bmatrix} 2 & 0 \\ 1 & 2 \end{bmatrix}$

(vii) $\begin{bmatrix} -3 & 0 \\ 0 & -2 \end{bmatrix}$ (viii) $\begin{bmatrix} 2 & 0 \\ 0 & 0 \end{bmatrix}$ (ix) $\begin{bmatrix} 0 & -3 \\ 3 & 0 \end{bmatrix}$

2. Classify the fixed points of the following linear systems:

(i) $\dfrac{dx_1}{dt} = 2x_1 + x_2$ $\dfrac{dx_2}{dt} = 3x_1 + 4x_2$

(ii) $\dfrac{dx_1}{dt} = 2x_1 + x_2$ $\dfrac{dx_2}{dt} = 3x_1$

(iii) $\dfrac{dx_1}{dt} = x_1 + 2x_2$ $\dfrac{dx_2}{dt} = -2x_1 + 5x_2$

(iv) $\dfrac{dx_1}{dt} = x_1 + 2x_2$ $\dfrac{dx_2}{dt} = 9x_1 + 4x_2$

(v) $\dfrac{dx_1}{dt} = -x_2$ $\dfrac{dx_2}{dt} = 4x_1 - 4x_2$

(vi) $\dfrac{dx_1}{dt} = -4x_2$ $\dfrac{dx_2}{dt} = 4x_1$

3. For each of the following systems, find the eigenvalues and eigenvectors of the matrix of coefficients. Use the analytical solution of the canonical system to find the general solution for **x**. Sketch some of the phase curves near the origin in the x_2–x_1 phase plane.

(i) $\dfrac{dx_1}{dt} = 2x_1 + x_2 \qquad \dfrac{dx_2}{dt} = 3x_2$

(ii) $\dfrac{dx_1}{dt} = 3x_1 + 4x_2 \qquad \dfrac{dx_2}{dt} = 2x_1 + x_2$

(iii) $\dfrac{dx_1}{dt} = x_1 - 5x_2 \qquad \dfrac{dx_2}{dt} = x_1 - x_2$

(iv) $\dfrac{dx_1}{dt} = 2x_1 + x_2 \qquad \dfrac{dx_2}{dt} = x_1 - x_2$

4. For each matrix **A** given below find the appropriate transformation matrix **P** that maps **A** to its canonical form and show that $\mathbf{P}^{-1}\mathbf{AP}$ is in the appropriate form.

(i) $\begin{bmatrix} 4 & 1 \\ -1 & 2 \end{bmatrix}$ (ii) $\begin{bmatrix} 2 & 2 \\ 3 & 1 \end{bmatrix}$ (iii) $\begin{bmatrix} 2 & 1 \\ 0 & 2 \end{bmatrix}$

(iv) $\begin{bmatrix} 1 & 2 \\ 1 & 1 \end{bmatrix}$ (v) $\begin{bmatrix} 1 & -2 \\ 2 & 3 \end{bmatrix}$ (vi) $\begin{bmatrix} 0 & -1 \\ 4 & 4 \end{bmatrix}$

5. Find the fixed points of the following systems and classify them.

(i) $\dfrac{dx_1}{dt} = x_1 + 3x_2 + 4 \qquad \dfrac{dx_2}{dt} = -6x_1 + 5x_2 - 1$

(ii) $\dfrac{dx_1}{dt} = x_1 + 3x_2 + 1 \qquad \dfrac{dx_2}{dt} = -6x_1 - 4x_2 + 1$

(iii) $\dfrac{dx_1}{dt} = 3x_1 + x_2 + 1 \qquad \dfrac{dx_2}{dt} = -x_1 + x_2 - 6$

(iv) $\dfrac{dx_1}{dt} = 2x_1 + 2x_2 - 2 \qquad \dfrac{dx_2}{dt} = 3x_1 + x_2 - 9$

(v) $\dfrac{dx_1}{dt} = x_1 + 2x_2 - 2 \qquad \dfrac{dx_2}{dt} = x_1 + x_2 - \dfrac{3}{2}$

3.5 Analysing non-linear systems

When given a non-linear velocity field **v**(**x**) one of the aims is to draw some of the phase portrait. The first step in the process is to find the positions of any fixed points. For a simple linear system there is at most one fixed point and the phase portrait is determined by its nature. In this section we investigate non-linear systems which are often much richer in detail and may contain several fixed points. The phase portraits are not always determined by the nature of the fixed points of the system.

For most non-linear systems we cannot solve the equations of motion explicitly; instead we investigate the local behaviour close to the fixed points, which are good approximations over part of the phase plane. This restriction is referred to as the **local phase portrait**. The complete or **global phase portrait** can often be built up from the local phase portraits.

We begin by determining the positions of the fixed points.

Example 5

Find the fixed points of the system

$$\frac{dx_1}{dt} = x_2 + x_1 x_2$$

$$\frac{dx_2}{dt} = x_1^2 - x_2^2 - 2x_1$$

Solution

The fixed points are given by solutions of the pair of equations

$$x_2 + x_1 x_2 = 0 \qquad (3.15)$$

$$x_1^2 - x_2^2 - 2x_1 = 0 \qquad (3.16)$$

From equation (3.15), either $x_2 = 0$ or $x_1 = -1$. Substitute for these values in equation (3.16).

When $x_2 = 0$, equation (3.16) gives $x_1^2 - 2x_1 = 0$ with solutions $x_1 = 0$ and $x_1 = 2$. Two fixed points are at (0,0) and (2,0).

When $x_1 = -1$, equation (3.16) gives $-x_2^2 + 3 = 0$ with solutions $x_2 = \pm\sqrt{3}$. Two fixed points are at $(-1, -\sqrt{3})$ and $(-1, \sqrt{3})$.

For this non-linear system there are four fixed points at (0,0), $(-1,-\sqrt{3})$, $(-1,\sqrt{3})$ and (2,0).

Figure 3.17 shows the direction field and some of the phase curves for the non-linear system of Example 5. We can see that the phase portrait is much richer in detail and that the nature of the four fixed points needs closer investigation.

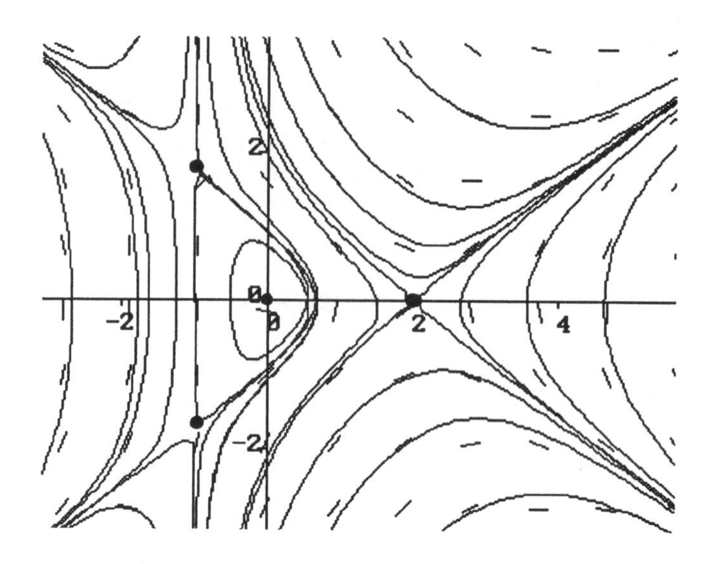

Fig 3.17 The direction field and some phase curves for the system $\dfrac{dx_2}{dx_1} = \dfrac{x_1^2 - x_2^2 - 2x_1}{x_2 + x_1x_2}$

TUTORIAL PROBLEM 3.10

Find the fixed points of the following system:

$$\frac{dx_1}{dt} = -x_1^2 + x_2 + 2$$

$$\frac{dx_2}{dt} = x_2^2 - x_1^2$$

In order to study the local phase portrait near a fixed point $\mathbf{x}_f$ we linearise the equations of motion using a Taylor series expansion. This is a similar approach to that adopted for the first order systems in Chapter 2. An important theory that justifies the approach is **the linearization theorem** which relates the local phase portrait of a non-linear system in the neighbourhood of a simple fixed point to that of the linearized system. The proof of the theorem is beyond the scope of this book.

Linearization theorem

Let the non-linear system

$$\frac{d\mathbf{x}}{dt} = \mathbf{v}(\mathbf{x})$$

have a simple fixed point at $\mathbf{x}_f$, then in a neighbourhood of $\mathbf{x}_f$ the phase portraits of the system and its linearization are qualitatively equivalent provided that the fixed point of the linearized system is not a centre. In particular, the fixed point $\mathbf{x}_f$ of the non-linear system is strongly stable whenever the origin is a strongly stable fixed point of the linear system, and $\mathbf{x}_f$ is unstable whenever the origin is unstable for the linearized system.

This important theorem provides the basis for our method of approach of analysing the stability of the fixed points of non-linear systems. The one case when the qualitative comparison breaks down is if the linearized system has a centre at the origin. The non-linear system could have a spiral or centre at $\mathbf{x}_f$. Example 6 illustrates the problem.

Example 6

Show that for each of the following non-linear systems the origin is a centre for each of the linearized systems.

(i) $$\frac{dx_1}{dt} = x_2 - x_1(x_1^2 + x_2^2), \qquad \frac{dx_2}{dt} = -x_1 - x_2(x_1^2 + x_2^2)$$

(ii) $$\frac{dx_1}{dt} = x_2 + x_1(x_1^2 + x_2^2), \qquad \frac{dx_2}{dt} = -x_1 + x_2(x_1^2 + x_2^2)$$

Use polar coordinates to find an analytical solution for each system showing that the linearized systems have qualitatively different phase portraits.

Solution

It is clear that the origin (0,0) is a fixed point for each system because $\mathbf{v}(0,0) = \mathbf{0}$. The linear approximation for each system about (0,0) is obtained by omitting the non-linear terms

$$\frac{dx_1}{dt} = x_2$$

$$\frac{dx_2}{dt} = -x_1$$

We can solve these equations by eliminating t and solving the resulting variables separable differential equation

$$\frac{dx_2}{dx_1} = -\frac{x_1}{x_2}$$

$$\Rightarrow \qquad x_1^2 + x_2^2 = c$$

This equation represents circles so that the linearized system has a centre, a stable fixed point, at the origin.

Use polar coordinates

$$x_1 = r\cos\theta \quad \text{and} \quad x_2 = r\sin\theta$$

$$\frac{dx_1}{dt} = \cos\theta\frac{dr}{dt} - r\sin\theta\frac{d\theta}{dt} \quad \text{and} \quad \frac{dx_2}{dt} = \sin\theta\frac{dr}{dt} + r\cos\theta\frac{d\theta}{dt}$$

On substituting into the differential equations and simplifying (the analysis is very similar to that for equations (3.13) and (3.14) on page 74) we obtain the following equations for each system:

(i) $\quad \dfrac{dr}{dt} = -r^3, \qquad \dfrac{d\theta}{dt} = -1$

(ii) $\quad \dfrac{dr}{dt} = r^3, \qquad \dfrac{d\theta}{dt} = -1$

Solving each pair of equations and eliminating t gives

(i) $\quad r = \dfrac{1}{\sqrt{\theta_0 - \theta}} \qquad$ (ii) $\quad r = \dfrac{1}{\sqrt{\theta_0 + \theta}}$

In each case the phase portrait is a spiral. For system (i), since $\dfrac{dr}{dt} < 0$ for all $r > 0$, the phase curves spiral inwards leading to a strongly stable fixed point at the origin. On the other hand for system (ii), $\dfrac{dr}{dt} > 0$ for all $r > 0$, so the phase curves spiral outwards leading to a strongly unstable fixed point at the origin. Figure 3.18 shows the qualitatively different behaviour of each system and its linearization.

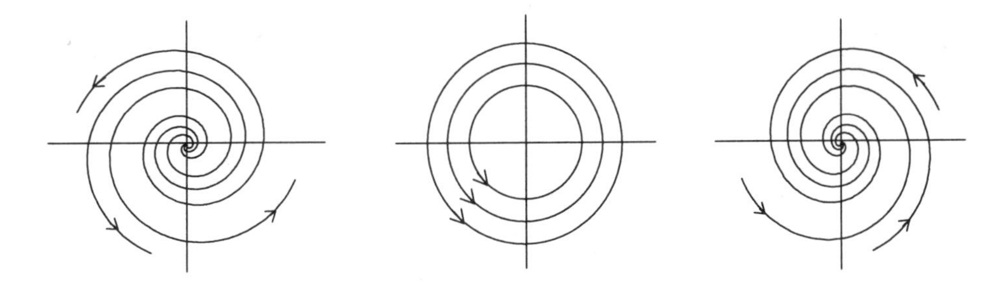

Fig 3.18 Phase portraits for systems (i), (ii) and their linearization

For the non-linear systems in Example 6 the linearized system was easily found by omitting the obvious non-linear terms. In most problems we linearize the equations of motion by using a Taylor expansion around the fixed point $\mathbf{x}_f$. Consider the non-linear velocity field $\mathbf{v}(\mathbf{x})$ with fixed point $\mathbf{x}_f = (\alpha,\beta)$. The first order Taylor expansion of $\mathbf{v}$ about (α,β) is

$$v_1(\mathbf{x}) = v_1(\mathbf{x}_f) + (x_1 - \alpha)\frac{\partial v_1}{\partial x_1}(\mathbf{x}_f) + (x_2 - \beta)\frac{\partial v_1}{\partial x_2}(\mathbf{x}_f)$$

$$v_2(\mathbf{x}) = v_2(\mathbf{x}_f) + (x_1 - \alpha)\frac{\partial v_2}{\partial x_1}(\mathbf{x}_f) + (x_2 - \beta)\frac{\partial v_2}{\partial x_2}(\mathbf{x}_f)$$

where each of the first order partial derivatives is evaluated at the fixed point. Since $\mathbf{x}_f$ is a fixed point, $v_1(\mathbf{x}_f) = 0$ and $v_2(\mathbf{x}_f) = 0$. Furthermore the fixed point can be moved to the origin by choosing new variables $\mathbf{X} = \mathbf{x} - \mathbf{x}_f$. In the $\mathbf{X}$ coordinate system $\mathbf{X}_f = (0,0)$.

The linearized system in the $\mathbf{X}$ phase space is

$$\frac{d\mathbf{X}}{dt} = \mathbf{A}\mathbf{X}$$

where $\mathbf{A}$ is the matrix of first order partial derivatives evaluated at the point (α,β).

$$\mathbf{A} = \begin{bmatrix} \frac{\partial v_1}{\partial x_1} & \frac{\partial v_1}{\partial x_2} \\ \frac{\partial v_2}{\partial x_1} & \frac{\partial v_2}{\partial x_2} \end{bmatrix}$$

The linearized system has an isolated fixed point provided that the matrix $\mathbf{A}$ is non-singular. We then say that the original non-linear system has a **simple** fixed point at $\mathbf{x}_f$.

A fixed point of a non-linear system is said to be **non-simple** if the corresponding linearized system is non-simple. In the analysis of linear systems we saw that non-simple linear systems contain a straight line, or sometimes a whole plane, of fixed points. Similarly lines of fixed points can also occur in the phase portrait of non-linear systems, however these lines are not necessarily straight.

In this book we will only investigate non-linear systems which have simple fixed points. The following example shows the method of analysing such systems by finding the matrix **A** and using the linearization theorem.

Example 7

Investigate the nature of the fixed points of the system of Example 5

$$\frac{dx_1}{dt} = x_2 + x_1x_2$$

$$\frac{dx_2}{dt} = x_1^2 - x_2^2 - 2x_1$$

which were found to be (0,0), $(-1,-\sqrt{3})$, $(-1,\sqrt{3})$ and (2,0).

Solution

The matrix of partial derivatives of the velocity field is

$$\begin{bmatrix} \dfrac{\partial v_1}{\partial x_1} & \dfrac{\partial v_1}{\partial x_2} \\ \dfrac{\partial v_2}{\partial x_1} & \dfrac{\partial v_2}{\partial x_2} \end{bmatrix} = \begin{bmatrix} x_2 & 1+x_1 \\ 2x_1 - 2 & -2x_2 \end{bmatrix}$$

At the origin (0,0)

$$\mathbf{A} = \begin{bmatrix} 0 & 1 \\ -2 & 0 \end{bmatrix}$$

Since tr(**A**) = 0 and det(**A**) = 2, Fig 3.15 shows that the fixed point of the linearized system is a centre. The linearization theorem tells us that the origin could be a centre or a spiral.

At (2,0)

$$\mathbf{A} = \begin{bmatrix} 0 & 3 \\ 2 & 0 \end{bmatrix}$$

Since tr(**A**) = 0 and det(**A**) = –6 the fixed point of the linear system is a saddle and the point (2,0) is therefore a saddle for the non-linear system.

At $(-1,\sqrt{3})$

$$\mathbf{A} = \begin{bmatrix} \sqrt{3} & 0 \\ -4 & -2\sqrt{3} \end{bmatrix} \quad \Rightarrow \quad \mathrm{tr}(\mathbf{A}) = -\sqrt{3},\ \det(\mathbf{A}) = -6$$

Since $(\mathrm{tr}(\mathbf{A}))^2 - 4\det(\mathbf{A}) = 27 > 0$ and $\mathrm{tr}(\mathbf{A}) < 0$ the fixed point of the non-linear system at $(-1,\sqrt{3})$ is a stable node.

At $(-1,-\sqrt{3})$

$$\mathbf{A} = \begin{bmatrix} -\sqrt{3} & 0 \\ -4 & 2\sqrt{3} \end{bmatrix} \quad \Rightarrow \quad \mathrm{tr}(\mathbf{A}) = \sqrt{3},\ \det(\mathbf{A}) = -6$$

The fixed point of the non-linear system is an unstable node.

TUTORIAL PROBLEM 3.11

Investigate the nature of the fixed points of the system in Tutorial Problem 3.10 on page 82.

The next example shows how the phase portrait of the linear system may only apply in a small neighbourhood of the fixed point.

Example 8

For the following non-linear system:

(i) find the position of the fixed point(s),

(ii) investigate the nature of the fixed point(s).

$$\frac{dx_1}{dt} = x_1 - x_2^3$$

$$\frac{dx_2}{dt} = x_2 + x_1^3$$

Compare the phase portraits of the non-linear system and the linear system in the region $-2 < x_1 < 2$, $-2 < x_2 < 2$.

Solution

(i) The system has only one fixed point at the origin (0,0).

(ii) The matrix of partial derivatives of the velocity field is

$$\begin{bmatrix} 1 & -3x_2^2 \\ 3x_1^2 & 1 \end{bmatrix}$$

At the origin (0,0)

$$\mathbf{A} = \begin{bmatrix} 1 & 0 \\ 0 & 1 \end{bmatrix} \quad \Rightarrow \quad \text{tr}(\mathbf{A}) = 2, \ \det(\mathbf{A}) = 1$$

Since $(\text{tr}(\mathbf{A}))^2 - 4\det(\mathbf{A}) = 0$ and $\text{tr}(\mathbf{A}) > 0$ the fixed point of the linearized system is an unstable star.

Figure 3.19 shows the direction field and some phase curves of the non-linear system and the unstable star. Close to the origin the phase curves are star-like straight lines but they soon curve; however the unstable star consists of straight lines over the whole region.

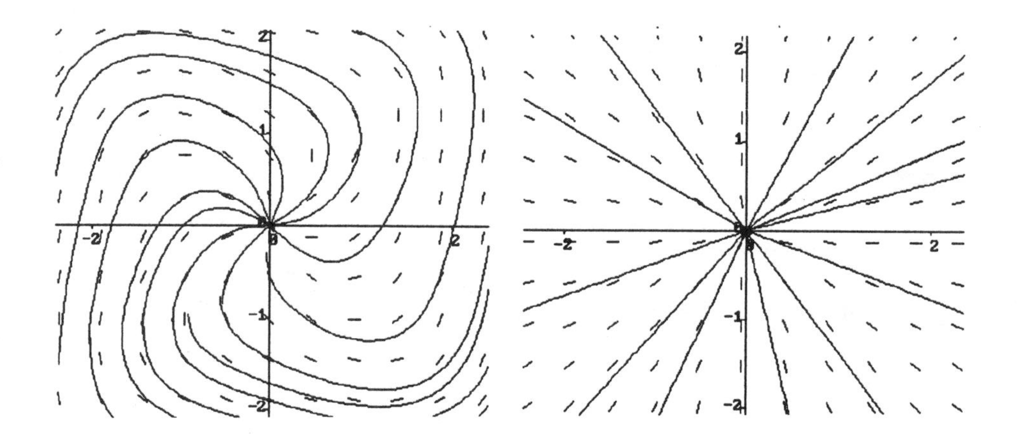

Fig 3.19 Local phase portrait for the non-linear system $\frac{dx_1}{dt} = x_1 - x_2^3, \ \frac{dx_2}{dt} = x_2 + x_1^3$ and its linearized system

This example shows clearly the localized behaviour of the phase portrait in a neighbourhood of the fixed point. For a system with several fixed points we can attempt to build up a global picture of the phase portrait from the localized pictures. However, the next tutorial problem shows that care is needed.

TUTORIAL PROBLEM 3.12

Show that the fixed points for linearized system of the non-linear system

$$\frac{dx_1}{dt} = x_1(2 - x_1)$$

$$\frac{dx_2}{dt} = -x_2(1 - x_1)$$

are saddle points.

Figure 3.20 is proposed as a phase portrait for this system which is consistent with the local behaviour near each fixed point. Investigate whether this is the correct global phase portrait.

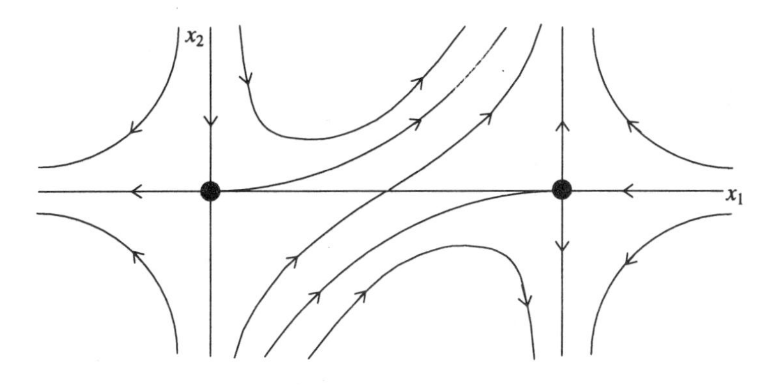

Fig 3.20 Proposed local phase portrait

This tutorial problem shows that care is needed when deducing a global phase portrait from local phase portraits close to fixed points. We often need to look for extra information about the phase curves that cannot be provided by the linearized system. In this problem we were able to observe that the lines $x_1 = 0$, $x_1 = 2$ and $x_2 = 0$ are phase curves. Since phase curves cannot cross, the phase curve in Fig 3.20 that crosses the x_1-axis at $x_1 = 1$ is clearly wrong.

Another feature of non-linear systems is illustrated in the next example. (This example is given in terms of polar coordinates (r,θ), and the solution of the system would be given as $r = r(\theta)$. The phase portrait would then be obtained using polar plots.)

Example 9

Investigate the phase portrait of the non-linear system with differential equation given in polar form

$$\frac{dr}{dt} = \alpha r(1-r)$$

$$\frac{d\theta}{dt} = 1$$

where α is a constant.

Solution

If we eliminate t and solve the variables separable differential equation we obtain the general solution

$$r = \frac{1}{1+ce^{-\alpha\theta}}$$

If $-1 < c < 0$ the motion occurs in the region $r > 1$, and for $c > 0$ the motion is restricted to the region $r < 1$.

There is a fixed point at the origin which is an unstable spiral. The interesting new feature is a closed circle $r = 1$ (which is given by $c = 0$). Figure 3.21 shows the phase portrait for this system, for $\alpha = 0.2$.

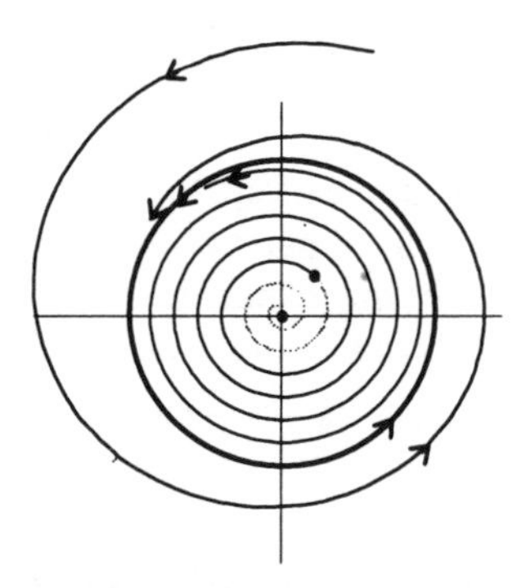

Fig 3.21 Phase portrait for $\frac{dr}{dt} = r(1-r)$, $\frac{d\theta}{dt} = 1$

The closed phase curve in Fig 3.21 is an example of a **limit cycle**. In this example of a limit cycle the phase curves are attracted towards it as time increases (time increasing $\Rightarrow$ polar angle θ increasing in this example because $\frac{d\theta}{dt} > 0$). This limit cycle is called an **attracting limit cycle**. Not all limit cycles have this property. Figure 3.22 shows a limit cycle which attracts the motion outside the cycle but has the nature of a centre inside the cycle.

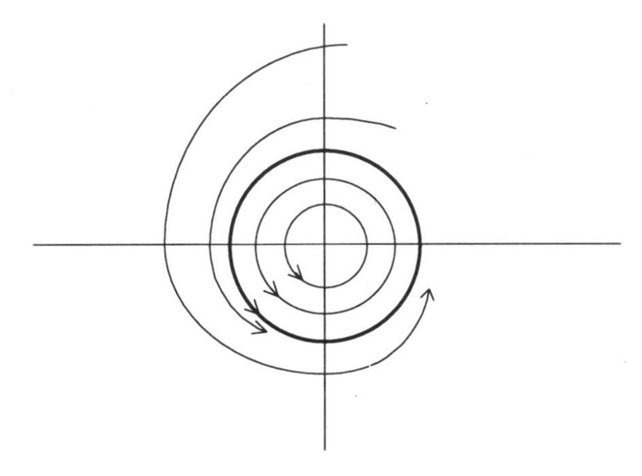

Fig 3.22 A limit cycle which is attracting

In the example above and Tutorial Problem 3.13 below, the limit cycle is a circle and can be analysed using polar coordinates. However, most limit cycles are not circular and identifying them is both difficult and important.

TUTORIAL PROBLEM 3.13

Use polar coordinates to find and analyse the limit cycles of the non-linear system

$$\frac{dx_1}{dt} = x_2 + x_1(1 - x_1^2 - x_2^2)$$

$$\frac{dx_2}{dt} = -x_1 + x_2(1 - x_1^2 - x_2^2)$$

Exercises 3.5

1. Find and classify the fixed points of each of the following systems:

(i) $\dfrac{dx_1}{dt} = -2x_2 + x_1x_2 - 8$ $\qquad \dfrac{dx_2}{dt} = x_2^2 - x_1^2$

(ii) $\dfrac{dx_1}{dt} = x_2 - x_1^2 + 2$ $\qquad \dfrac{dx_2}{dt} = x_1^2 - x_2^2$

(iii) $\dfrac{dx_1}{dt} = 1 + x_1x_2$ $\qquad \dfrac{dx_2}{dt} = x_2(1 + x_1)$

(iv) $\dfrac{dx_1}{dt} = x_2^2 - 3x_1 + 2$ $\qquad \dfrac{dx_2}{dt} = x_1^2 - x_2^2$

(v) $\dfrac{dx_1}{dt} = x_2 + x_1 - x_1 x_2$ $\qquad$ $\dfrac{dx_2}{dt} = x_1 + x_2 - x_2^2$

(vi) $\dfrac{dx_1}{dt} = x_1^2 - e^{x_2}$ $\qquad$ $\dfrac{dx_2}{dt} = x_2(1 + x_2)$

2. Transform each of the following second order differential equations into a pair of simultaneous first order equations. Classify the fixed points of each system.

(i) $\dfrac{d^2x}{dt^2} - \dfrac{dx}{dt} + x^2 - 2x = 0$

(ii) $\dfrac{d^2x}{dt^2} - \left(\dfrac{dx}{dt}\right)^3 + x + 5 = 0$

3. Sketch the phase portraits in the neighbourhood of each of the fixed points of the system

$$\frac{dx_1}{dt} = x_1(1 - x_1^2) \qquad \frac{dx_2}{dt} = x_2$$

and use your results to suggest a global phase portrait.

4. If one population of a competing pair of populations is harvested, the equilibrium structure can be radically changed. For instance, stable coexistence could become competitive exclusion or vice versa. In the following competition model, H is a constant harvest rate. Find the equilibrium points of the system for (i) $H = 0$, (ii) $H = \frac{1}{2}$; analyse their stability, and sketch the phase plane.

$$\frac{dx}{dt} = x\left(1 - \frac{1}{8}x - \frac{1}{10}y\right) - H$$

$$\frac{dy}{dt} = y\left(1 - \frac{1}{8}y - \frac{1}{12}x\right)$$

5. The equation of motion for a freely pivoted pendulum is

$$\frac{d^2x}{dt^2} + \omega^2 \sin x = 0$$

where ω is a constant and x is the angular displacement.

(i) Investigate the nature of the fixed points of this system. Use your results to suggest a global phase portrait.

(ii) Show that an analytical solution for this system is

$$y^2 - 2\omega^2 \cos x = c$$

where $\dfrac{dx}{dt} = y$ and c is a constant. Sketch graphs of this solution and compare them with your global phase portraits in part (i).

6. Classify the nature of the fixed point at the origin for the linearized system of the non-linear system

$$\frac{dx_1}{dt} = -x_2$$

$$\frac{dx_2}{dt} = x_1 - x_1^3$$

Show that the equation of the phase curves is

$$x_1^2 - \frac{x_1^4}{2} + x_2^2 = c$$

where c is a constant. Sketch these curves for various values of c and show that the local phase portrait of the non-linear system and its linearization are qualitatively equivalent. Why could this conclusion not be deduced directly from the linearization theorem?

7. Sketch the phase portraits for each of the following non-linear systems given in terms of polar coordinates:

(i) $\dfrac{dr}{dt} = \alpha r(r-1)^2, \qquad \dfrac{d\theta}{dt} = 1$ where α is a constant

(ii) $\dfrac{dr}{dt} = \begin{cases} r(2-r) & \text{if } r \le 2 \\ 0 & \text{otherwise} \end{cases}, \qquad \dfrac{d\theta}{dt} = 1$

In each case identify any limit cycles.

8. Using polar coordinates to transform the differential equations show that there exists a region $R = \{(x_1, x_2) | x_1^2 + x_2^2 \le r^2\}$ such that all trajectories of the system

$$\frac{dx_1}{dt} = -wx_2 + x_1(1 - x_1^2 - x_2^2), \qquad \frac{dx_2}{dt} = wx_1 + x_2(1 - x_1^2 - x_2^2) - F$$

where w and F are constants, eventually enter R. Show that the system has a limit cycle when $F = 0$.

3.6 Case studies

We now investigate two case studies which use the ideas of this chapter. The first is about the dangers of lead in the environment and the second extends the population modelling of Chapter 2.

Lead absorption in the body

Lead in the environment caused by pollution from motor vehicles has been identified as a major health hazard. Some lead is part of the natural environment and without human activities the concentration of lead in the atmosphere is so low that its effects on humans would be negligible. On the other hand, our industrial society pumps lead into the atmosphere and the major source is vehicle exhaust. The effects of lead poisoning are particularly bad for those living close to motorways with heavy traffic flows. Research has shown that people living in such areas can be freed of symptoms such as headache, fatigue, indigestion and nervous disturbance by anti-lead treatment, thus linking lead poisoning with traffic.

To model the build up of lead pollution in the body we use a simple compartmental model. We assume that lead enters the body through breathing and food and builds up in the blood, in body tissues and in the bones. Some of the lead is removed through urine, hair, nails, skin and sweat. From the blood lead is rapidly distributed in the tissues. The absorption by the bones is a much slower process. However when the body reaches an equilibrium state the skeleton contains 90% or more of the lead.

Because of the much longer timescale for damage to the bones we begin by modelling a system of blood and body tissue only. Suppose that a model of this system is formulated by tracking the concentrations of lead into and out of the blood and tissue. Figure 3.23 shows a two compartmental model where $x_1(t)$ and $x_2(t)$ are the amounts of lead in the blood and body tissue respectively at time t.

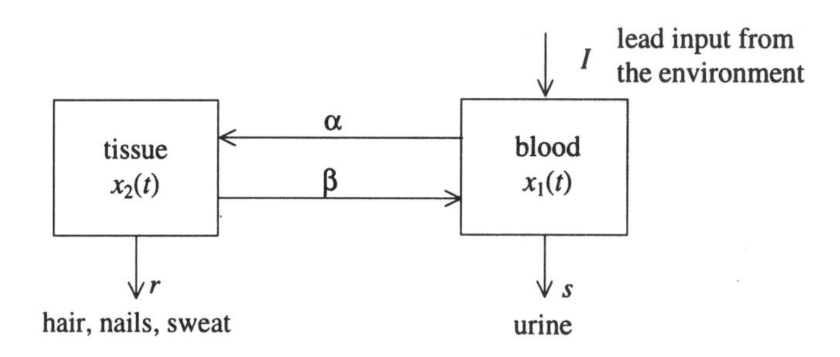

Fig 3.23 A two compartmental model for dynamics of lead absorption from the environment to the blood and body tissue

Let I be the daily input of lead (in μg per day) into the blood and body tissue directly from food, drink and breathing. The model will assume that the rate of absorption of lead from one compartment of the model to another is proportional to the amount of lead present in the compartment. For example, the rate of transfer of lead from blood to tissue is $\alpha x_1(t)$. The constants are α, β, r and s.

The basic input-output principle

accumulation = input – output

leads to the following pair of first order differential equations:

$$\frac{dx_1}{dt} = (I + \beta x_2) - (\alpha x_1 + s x_1) = -(s+\alpha)x_1 + \beta x_2 + I$$

$$\frac{dx_2}{dt} = \alpha x_1 - (\beta x_2 + r x_2) = \alpha x_1 - (\beta + r)x_2$$

This is a linear dynamical system. Using the techniques developed in this chapter we can analyse the equilibrium point.

Example 10

The following model of lead poisoning was formulated in the 1970s by a group of scientists who studied the lead intake and excretion of a healthy volunteer living in an area of heavy smog in Los Angeles, USA:

$$\frac{dx_1}{dt} = -0.0361x_1 + 0.0124x_2 + 49.3$$

$$\frac{dx_2}{dt} = 0.0111x_1 - 0.0286x_2$$

where the daily intake is 49.3 μg per day. Find the equilibrium level of lead in the blood and body tissue and classify the equilibrium point.

Solution

The equilibrium level (or fixed point) is given by

$$\begin{aligned} -0.0361x_1 + 0.0124x_2 + 49.3 &= 0 \\ 0.0111x_1 - 0.0286x_2 \qquad &= 0 \end{aligned}$$

Solving for x_1 and x_2 we get (to the nearest integer)

$$x_1 = 1576\ \mu\text{g} \quad x_2 = 612\ \mu\text{g}$$

The matrix of coefficients of the system is

$$\mathbf{A} = \begin{bmatrix} -0.0361 & 0.0124 \\ 0.0111 & -0.0286 \end{bmatrix}$$

and tr(**A**) = –0.0647 and det(**A**) = 0.000 894 819 so that

$$(\text{tr}(\mathbf{A}))^2 - 4\det(\mathbf{A}) = 0.000\,606\,813 > 0$$

The eigenvalues are real distinct and since tr(**A**) < 0 the fixed point is a stable node. Hence we deduce that the lead levels of 1576 μg in the blood and 612 μg in the body tissues are stable.

The values of the eigenvalues for matrix **A** are

$$\lambda_1 = -0.02 \quad \text{and} \quad \lambda_2 = -0.045$$

The general solutions for x_1 and x_2 are then given by

$$x_1 = 1576 + ae^{-0.02t} + be^{-0.045t}$$

$$x_2 = 612 + 1.3ae^{-0.02t} - 0.72e^{-0.045t}$$

where a and b are constants which depend on $x_1(0)$ and $x_2(0)$.

TUTORIAL PROBLEM 3.14

For the model in Example 10 assume that the absorption and excretion rates α, β, r and s remain the same but the volunteer lives in a rural environment where the daily intake of lead is 30 μg, mostly from food, with very little pollution from the atmosphere. In a sense this is the unavoidable lead intake.

Find and classify the new equilibrium level. Suppose now that the lead intake from the environment is doubled so that the total daily intake of lead is 70 μg. Find the new equilibrium level.

The stability of the system depends on the matrix **A** and in all cases the equilibrium levels are stable. This means that unless a person moves away from an environment of high lead pollution the long term effects will be at the constant equilibrium levels. The only way to reduce the dangers of lead poisoning is to cut the lead input into the environment.

TUTORIAL PROBLEM 3.15

The real danger for the human body is the absorption of lead by the bones. This introduces a third compartment to our model.

(i) Extend the model formulated above to include a third compartment, the bones. In your model let the amount of lead in the bones be x_3 and assume that the lead is transferred to and from the bones from the blood only.

(ii) The following model of lead poisoning was formulated in the 1970s by a group of scientists who studied the lead intake and excretion of a healthy volunteer living in an area of heavy smog in Los Angeles, USA:

$$\text{(blood)} \qquad \frac{dx_1}{dt} = -0.0361x_1 + 0.0124x_2 + 0.000\,035x_3 + 49.3$$

$$\text{(body tissue)} \qquad \frac{dx_2}{dt} = 0.0111x_1 - 0.0286x_2$$

$$\text{(bones)} \qquad \frac{dx_3}{dt} = 0.0039x_1 - 0.000\,035x_3$$

In these equations the quantities of lead and time are measured in μg and days respectively. The ingestion rate is 49.3 μg per day. Find the equilibrium point for this system giving your answer in terms of the lead content in the bones, blood and body tissue measured in grams. Show that the eigenvalues are negative so that the system is stable.

Interacting species

In the case study in Chapter 2 we introduced the logistic model for describing the population growth of a single isolated species

$$\frac{dP}{dt} = aP\left(1 - \frac{P}{M}\right)$$

where P is the population, M is the equilibrium population and a is a constant low population growth rate (see page 39 for the formulation of this model).

Suppose that we add a second species of predators which uses the first species as its main food source. Let the population of prey and predator be denoted by $P_1(t)$ and $P_2(t)$ respectively. We add to the assumption in formulating the logistic model in Chapter 2 by assuming that the number of predator–prey interactions during each time interval is proportional to $P_1(t)$ and $P_2(t)$. This leads to a simple model that the change in population of predator and prey is proportional to the product P_1P_2.

Introducing this into the logistic model gives the predator–prey equations

$$\text{Prey} \qquad \frac{dP_1}{dt} = P_1(\alpha - \beta P_1 - rP_2)$$

$$\text{Predator} \qquad \frac{dP_2}{dt} = P_2(-\gamma - \delta P_2 + sP_1) \tag{3.17}$$

where α, β, γ, δ, r and s are positive constants. (In the second equation P_2 must become zero if there is no prey P_1, which is the reason for writing $-\gamma P_2$ for small population numbers.) There are several possible outcomes from the predator–prey iteration:

- predator dies out and prey survives to equilibrium;
- prey dies out and then predator dies out;
- the two species coexist in equilibrium.

Each of the outcomes can be investigated using the analysis of the fixed points of the system (3.17). A necessary condition for equilibrium of the two species is a fixed point such that $P_1 > 0$ and $P_2 > 0$. To find the fixed points we solve

$$P_1(\alpha - \beta P_1 - rP_2) = 0 \tag{3.18a}$$

$$P_2(-\gamma - \delta P_2 + sP_1) = 0 \tag{3.18b}$$

From equation (3.18a)

$$P_1 = 0 \quad \text{or} \quad \alpha - \beta P_1 - rP_2 = 0$$

From equation (3.18b)

$$P_2 = 0 \quad \text{or} \quad -\gamma - \delta P_2 + sP_1 = 0$$

Hence (0,0) is a fixed point in which both species become extinct. Also for $P_1 = 0$, we have $P_2 = -\dfrac{\gamma}{\delta} < 0$ which is not physically realistic. Also for $P_2 = 0$, we have $P_1 = \dfrac{\alpha}{\beta} > 0$. Hence $\left(\dfrac{\alpha}{\beta}, 0\right)$ is a fixed point in which the predator is extinct and the prey survives at the equilibrium level $\dfrac{\alpha}{\beta}$.

A fourth fixed point is $\left(\dfrac{\alpha\delta + \gamma r}{\beta\delta + rs}, \dfrac{\alpha s - \beta\gamma}{\beta\delta + rs}\right)$ representing both species surviving in equilibrium provided $\alpha s > \beta\gamma$.

To classify the fixed points we find the matrix for the linearized system evaluated at each point.

$$\mathbf{A} = \begin{bmatrix} \alpha - 2\beta P_1 - rP_2 & -rP_1 \\ sP_2 & -\gamma - 2\delta P_2 + sP_1 \end{bmatrix}$$

At (0,0) we have $\mathbf{A} = \begin{bmatrix} \alpha & 0 \\ 0 & -\gamma \end{bmatrix}$ and the fixed point is an (unstable) saddle point.

At $\left(\frac{\alpha}{\beta}, 0\right)$ we have $\mathbf{A} = \begin{bmatrix} -\alpha & -\frac{r\alpha}{\beta} \\ 0 & -\gamma + \frac{\alpha s}{\beta} \end{bmatrix}$ and the fixed point is a stable node provided $s < \frac{\beta\gamma}{\alpha}$.

At $\left(\frac{\alpha\delta + \gamma r}{\beta\delta + rs}, \frac{\alpha s - \beta\gamma}{\beta\delta + rs}\right)$ we have $\mathbf{A} = \frac{1}{\beta\delta + rs}\begin{bmatrix} -\beta(\alpha\delta + r\gamma) & -r(\alpha\delta + \gamma r) \\ s(\alpha s - \beta\gamma) & -\delta(\alpha s - \beta\gamma) \end{bmatrix}$.

Now $\det(\mathbf{A}) = \frac{(\alpha\delta + \gamma r)(\alpha s - \beta\gamma)}{(\beta\delta + rs)} < 0$ if $s < \frac{\beta\gamma}{\alpha}$ and the fixed point is a saddle point.

If $s > \frac{\beta\gamma}{\alpha}$, the fixed point is a stable node provided $(\mathrm{tr}(\mathbf{A}))^2 - 4\det(\mathbf{A}) \geq 0$. Otherwise the fixed point is a centre and according to the linearization theorem we cannot classify the fixed point using the linearized matrix.

From this algebraic analysis we can deduce that if $s < \frac{\beta\gamma}{\alpha}$ then coexistence of the two species is extremely unlikely since the stable equilibrium condition is $P_1 = \frac{\alpha}{\beta}$ and $P_2 = 0$. For $s > \frac{\beta\gamma}{\alpha}$ the outcome is not so straightforward to predict. The fixed point at $\left(\frac{\alpha}{\beta}, 0\right)$ becomes unstable; however coexistence of species is now possible with $P_1 = \frac{\alpha\delta + \gamma r}{\beta\delta + rs}$ and $P_2 = \frac{\alpha s - \beta\gamma}{\beta\delta + rs}$ being a stable node, a centre or an attracting spiral.

TUTORIAL PROBLEM 3.16

Consider the predator–prey model

$$\frac{dP_1}{dt} = (1 - aP_1 - P_2)P_1$$

$$\frac{dP_2}{dt} = (-1 - aP_2 + P_1)P_2$$

where a is an positive parameter such that $0 \le a < 1$.

Show that, at the non-trivial fixed point, there is a centre for $a = 0$ which changes into a stable point for $0 < a < 1$.

Sketch the phase portrait for various values of a.

Find the range of values of a for the non-trivial fixed point to be a spiral.

Exercises 3.6

1. In this problem you model the spread of pollution in three lakes caused by an industrial plant that leaks chemicals into a river feeding one of the lakes. Figure 3.24 shows the three lakes are connected by small rivers and streams. The direction of flow between the lakes is shown by the arrows.

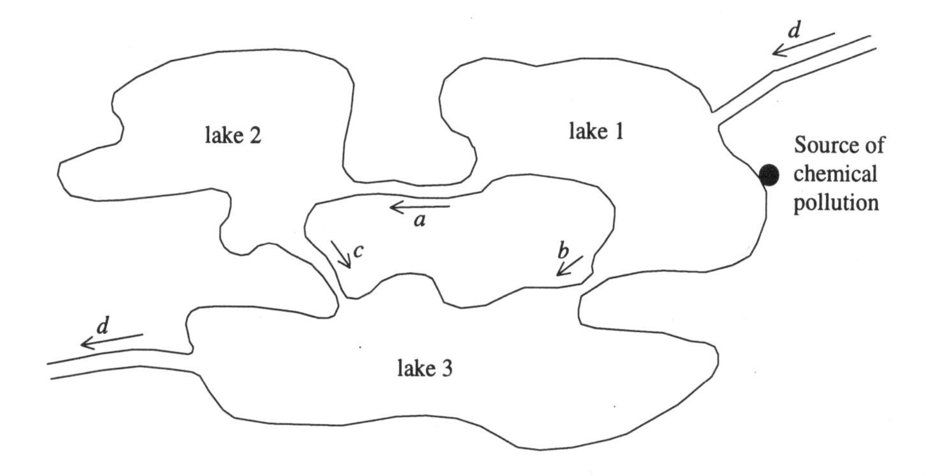

Fig 3.24 Pollution of three lakes by an industrial plant

A chemical is leaked into lake 1 at a rate $I(t)$ kg/year. The following table defines the variables and flow parameters shown in the figure.

Symbol	Units	Meaning
p_1, p_2, p_3	kg	Amount of pollutant in lake at time t
a, b, c	km^3/year	Flow rates between lakes
c_1, c_2, c_3	kg/km^3	Concentration of pollutant in lake at time t
V_1, V_2, V_3	km^3	Volume of water in each lake
d	km^3/year	Flow rate out of lake 3 (= flow rate into lake 1)

You can assume that the volume of each lake remains constant and that perfect mixing occurs within each lake.

Formulate a mathematical model in the form of three differential equations that describes the flow of pollutant through the system of lakes.

Consider the model for polluting the system of lakes which is applied to three of the Great Lakes system in North America. The data for each lake is shown in the table below.

Characteristic	Lake Michigan	Lake Erie	Lake Ontario
Volume of water (km^3)	4871	458	1636
Average outflow rates (km^3/year)	209	175	209
Industrial input of chemical pollution (kg/year)	50 000		

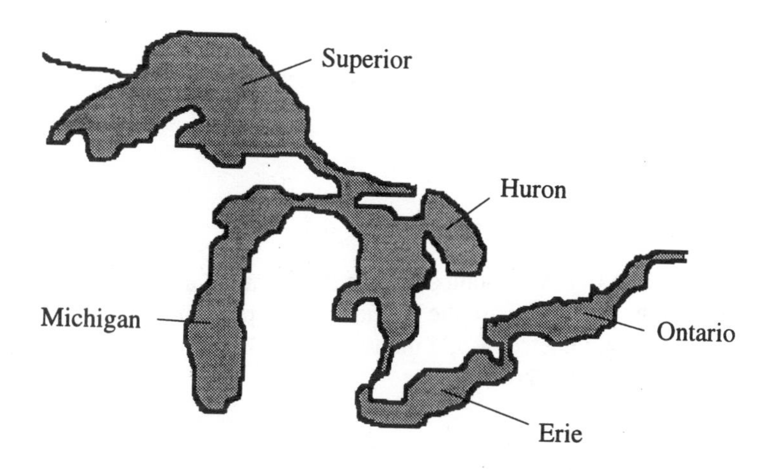

Fig 3.25 Map of great lakes

Assume that water flows directly from Lake Michigan into Lakes Erie and Ontario and directly from Lake Erie into Lake Ontario. Water leaves Lake Ontario into the River Hudson. Assume also that perfect mixing occurs.

Set up a mathematical model for the flow of pollution through the three Great Lakes and find the equilibrium pollution levels. Are the equilibrium levels stable?

2. A simple model for competition of species is where one population grows logistically and the other exponentially in the absence of competition. This leads to the model

$$\frac{dP_1}{dt} = P_1 a\left(1 - \frac{P_1}{b}\right) - \alpha P_1 P_2, \qquad \frac{dP_2}{dt} = cP_2 - \beta P_1 P_2$$

where a, b, c, α and β are all positive numbers. Show that in this case there is no possibility of stable coexistence of the two species.

3. If one population of a competing pair of populations is harvested, the equilibrium structure can be radically changed. For instance, stable coexistence could become competitive exclusion or vice versa. In the following competition models, H is a constant harvest rate. Find the equilibrium points of the system for each given value of H, analyse their stability, and sketch the phase plane.

$$\frac{dP_1}{dt} = P_1\left(1 - \frac{P_1}{10} - \frac{P_2}{10}\right) - H$$

$$\frac{dP_2}{dt} = P_2\left(1 - \frac{P_2}{5} - \frac{P_1}{8}\right)$$

$$H = 0, \frac{5}{4}, \frac{25}{4}$$

4. The system of differential equations

$$\frac{dS}{dt} = -rIS$$

$$\frac{dI}{dt} = rIS - cI$$

$$R = 1 - S - I$$

where r, c (> 0) are constants, models the spread of a disease through a population. S, I and R are the fractions of the population which are, respectively, susceptible to, infected by and immune to the disease. Assume the initial values S_0, I_0 and $R_0 = 0$. Show that

(i) if $\frac{rS_0}{c} \le 1$, then I decreases to zero,

(ii) if $\frac{rS_0}{c} > 1$, then I increases to a maximum value, and then decreases to zero. Find this maximum value.

Show also that in both cases (i) and (ii) the population of susceptibles S tends to the limit S_L as $t \to \infty$, where S_L is the unique solution of the equation

$$x = \frac{c}{r}\ln\left(\frac{x}{S_0}\right) + 1$$

in the interval $0 < x < \frac{c}{r}$.

Further exercises

1. What is a separable system? Show that the system

$$\frac{dx_1}{dt} = -5x_1 + 12x_2 \qquad \frac{dx_2}{dt} = 12x_1 + 5x_2$$

is separable in the variables (X_1, X_2) where $x_1 = 2X_1 + 3X_2$, $x_2 = 3X_1 - 2X_2$. Use the new variables as coordinates and construct a phase portrait for the system.

2. Suppose that a simple linear system has a matrix of coefficients **A** which is symmetric so that

$$\frac{d\mathbf{x}}{dt} = \mathbf{Ax} \quad \text{and} \quad \mathbf{A} = \begin{bmatrix} a & b \\ b & d \end{bmatrix}$$

Investigate the nature of the isolated fixed point that such a system can have.

3. The simple linear system

$$\frac{d\mathbf{x}}{dt} = \mathbf{Ax}$$

has a matrix of coefficients **A** of the form

$$\mathbf{A} = \begin{bmatrix} a & b \\ -b & d \end{bmatrix}$$

where $a > 0$ and $d > 0$. Investigate the nature of the isolated fixed point that such a system can have.

4. Classify the fixed point of each of the following linear systems:

(i) $$\frac{dx_1}{dt} = 2x_1 - 2x_2 \qquad \frac{dx_2}{dt} = -2x_1 + 2x_2$$

(ii) $\dfrac{dx_1}{dt} = -2x_1 - 2x_2$ $\qquad$ $\dfrac{dx_2}{dt} = -x_1 - 3x_2$

(iii) $\dfrac{dx_1}{dt} = x_1 + 5x_2$ $\qquad$ $\dfrac{dx_2}{dt} = -2x_1 + 3x_2$

(iv) $\dfrac{dx_1}{dt} = 4x_1 + 3x_2$ $\qquad$ $\dfrac{dx_2}{dt} = x_1 + 2x_2$

5. For which simple fixed points does the linearized system always describe the local behaviour of the non-linear system from which it is derived?

6. Find all the fixed points of the following non-linear systems and investigate their nature:

(i) $\dfrac{dx_1}{dt} = x_1(1 - x_2)$ $\qquad$ $\dfrac{dx_2}{dt} = 1 - x_1 x_2$

(ii) $\dfrac{dx_1}{dt} = x_1^2 - x_2^2$ $\qquad$ $\dfrac{dx_2}{dt} = 2x_1 x_2 - 6x_1 - 8$

(iii) $\dfrac{dx_1}{dt} = 4 - 4x_1^2 - x_2^2$ $\qquad$ $\dfrac{dx_2}{dt} = 3x_1 x_2$

(iv) $\dfrac{dx_1}{dt} = 1 - x_2^2$ $\qquad$ $\dfrac{dx_2}{dt} = x_1 e^{x_1}$

(v) $\dfrac{dx_1}{dt} = x_2 - x_1 - x_1^2$ $\qquad$ $\dfrac{dx_2}{dt} = x_2 - x_1 + x_1 x_2$

7. Consider the non-linear system

$$\frac{dx}{dt} = y \quad \text{and} \quad \frac{dy}{dt} = v(x, y)$$

where the velocity component $v(x,y)$ is an even function of y. Show that if the system has fixed points then they are either centres or saddle points.

8. Find the fixed points, and their nature, of the non-linear system

$$\frac{dx}{dt} = (1 + x - 2y)x \qquad \frac{dy}{dt} = (x - 1)y$$

By sketching the local phase portraits near each fixed point suggest a global phase portrait.

9. Consider the second order differential equation

$$\frac{d^2x}{dt^2} + x + x^3 = 0$$

(i) Convert it to a system of simultaneous first-order equations.

(ii) Show that the corresponding dynamical system has a single fixed point whose associated linearized system has a centre.

(iii) Show that

$$\left(\frac{dx}{dt}\right)^2 + x^2 + \frac{1}{2}x^4$$

remains constant along any phase curve of the system and deduce the nature of the fixed point for the non-linear system.

10. A damped vertical pendulum has the equation of motion

$$\frac{d^2\theta}{dt^2} + \alpha\frac{d\theta}{dt} + \omega^2 \sin\theta = 0 \qquad \alpha \geq 0$$

where θ is the angle between the pendulum and the downward vertical. Determine the position and the nature of the fixed points. Sketch the phase curves in a sufficiently large region of phase space to show all the qualitative features of the motion. Describe what happens to the motion and fixed points when $\alpha = 0$.

11. The following pair of differential equations is proposed as an economic model for the price P of a single item in a market where Q is the quantity of the item available:

$$\frac{dP}{dt} = aP\left(\frac{b}{Q} - P\right)$$

$$\frac{dQ}{dt} = cQ(fP - Q)$$

where a, b, c and f are positive constants.

(i) Consider the case $a = 1$, $b = 20\,000$, $c = 1$ and $f = 30$. Find and investigate the nature of the fixed points of this system. What can you say about the long term behaviour of P and Q?

(ii) Discuss this model suggesting on what assumptions and simplifications it is based.

12. The behaviour of a simple disc dynamo is governed by the system

$$\frac{dx}{dt} = -ax + xy$$

$$\frac{dy}{dt} = 1 - by - x^2$$

where a and b are positive parameters, x is the output current of the dynamo and y is the angular velocity of the rotating disc. Show that for $ab > 1$ there is one stable fixed point A at $(0, b^{-1})$, but for $ab < 1$ the fixed point A becomes a saddle point and stable fixed points occur at $(\pm\sqrt{1-ab}, a)$.

13. The following pair of differential equations is a simple predator–prey model for the population of foxes and rabbits in a park in Reading:

$$\frac{dF}{dt} = F(80 - 0.1R + 2F)$$

$$\frac{dR}{dt} = R(120 - 0.1R - 2F)$$

Find the equilibrium populations of foxes and rabbits and describe what happens to the number of foxes and rabbits for different initial populations.

14. Show that the fixed point at the origin for the linearized system of the non-linear system

$$\frac{dx_1}{dt} = -x_2$$

$$\frac{dx_2}{dt} = x_1 - x_1^5$$

is a centre.

Show that the equation of the phase curves is

$$x_1^2 - \frac{x_1^6}{3} + x_2^2 = c$$

where c is a constant. Sketch these curves for various values of c and show that the local phase portrait of the non-linear system and its linearization are qualitatively equivalent. Why could this conclusion not be deduced directly from the linearization theorem?

15. Consider the non-linear system

$$\frac{dx_1}{dt} = x_1^2 - x_2^3$$

$$\frac{dx_2}{dt} = x_1^2 (x_1^2 - x_2^3)$$

(i) Show that the non-linear system has a line of fixed points.

(ii) Show that every fixed point on the line is non-simple.

(iii) Investigate what conclusions you get by using the linearization theorem.

16. Consider the non-linear system

$$\frac{dx}{dt} = -y + xr^2 \sin\frac{\pi}{r}$$

$$\frac{dy}{dt} = x + yr^2 \sin\frac{\pi}{r}$$

(i) Show that the only simple fixed point is the origin (0,0).

(ii) Use polar coordinates $x = r\cos\theta$ and $y = r\sin\theta$ to show that the family of circles

$$c_n\colon\ r = \frac{1}{n} \qquad n = 1, 2, 3, \ldots$$

are closed trajectories.

(iii) Investigate the behaviour of the system between any two circles c_n and c_{n+1}.

4 •Discrete Systems

We now turn our attention to systems in which the models are difference equations (often called recurrence relations). Because the dependent variable is found at discrete values of the independent variable, these are called **discrete models**. The differential equation models of Chapters 2 and 3 are continuous models. Difference and differential equations approximate each other and sometimes one type of model is more amenable to solve than the other. Much of the terminology carries over from Chapters 2 and 3.

4.1 Examples of discrete systems

The equation

$$x_{n+1} = 0.2x_n + 2$$

is an example of a **difference equation**. It is an equation that is used iteratively to generate a sequence of numbers once the initial term is specified. Difference equations occur naturally as models for many discrete systems. The following example illustrates a simple application.

Example 1

Andrew, aged 18½, won £20 000 on the National Lottery. He invested it in the CTM Building Society at a fixed rate of interest of 6½% compounded annually. How much would Andrew have on his 25th birthday?

Solution

Let x_n be the money in pounds in the account at the end of the nth year. So at the end of the first year Andrew has

$$x_1 = \text{initial amount} + \text{interest}$$

$$x_1 = 20\,000 + \left(\frac{6\frac{1}{2}}{100} \times 20\,000\right) = 21\,300$$

At the end of the second year he has

$$x_2 = 21\,300 + \left(\frac{6\frac{1}{2}}{100} \times 21\,300\right) = 22\,684.5$$

and so on. This sequence can be written more generally as the difference equation

$$x_{n+1} = x_n + \frac{6\frac{1}{2}}{100}x_n$$

x_{n+1}	x_n	$\frac{6\frac{1}{2}}{100}x_n$
↑	↑	↑
amount at end of year $n+1$	amount at end of year n	interest

i.e. $x_{n+1} = 1.065x_n$

Andrew starts with £20 000 so $x_0 = 20\,000$. The difference equation generates the following sequence:

n	0	1	2	3	4	5	6
x_n	20 000	21 300	22 684.5	24 158.99	25 729.33	27 401.73	29 182.85

At the end of 6 years Andrew is aged 24½ and the amount has increased to £29 182.85. He becomes 25 before the next interest is added and so Andrew has £29 182.85 on his 25th birthday.

TUTORIAL PROBLEM 4.1

Investigate how much Andrew would have on his 25th birthday if interest was compounded at (i) 6 month intervals, (ii) monthly, and (iii) weekly.

Show that if the interest is compounded continuously then the discrete model for the amount after t years becomes a continuous exponential model.

Discrete models are often used in modelling population growth of species. In the case study in Chapter 2, we formulated a continuous differential equation model for the logistic model of population growth

$$\frac{dP}{dt} = aP\left(1 - \frac{P}{M}\right)$$

where a and M are constants. M is the equilibrium population level. One of the criticisms of this model is that it assumes continuous births of the species whereas in many cases the births occur once a year. (There are of course many other criticisms that we could make of this simple isolated species model.)

TUTORIAL PROBLEM 4.2

Let P_n be the fraction of the equilibrium population level of a species at generation n, so that $0 \le P_n \le 1$.

Formulate the difference equation logistic model

$$P_{n+1} = kP_n(1 - P_n)$$

Explain briefly what the constant k depends on and its relationship to a in the continuous model.

The difference equation for interest rates is an example of a linear equation, whereas the logistic population model is non-linear. We will see in this chapter that such a simple non-linear difference equation leads to very rich mathematical theory much of which is still not completely understood.

Exercises 4.1

1. On 1 January 1990 Elizabeth invested £2000 in a building society at a fixed interest rate of 5% compounded annually. At the end of each year (starting on 31 December 1990) she withdraws £150 to pay for Christmas presents. How much will be remaining in her account on 1 January 2000?

2. Mr and Mrs Bifurcation buy a new house with a mortgage of £30 000. This has to be paid back over 25 years at a rate of interest of 9½% compounded annually.

   Suppose that £M is the constant monthly repayment which ensures that the amount owed is zero after 25 years.

   (i) Obtain a formula for the amount owed after one year.

   (ii) Formulate a difference equation, in terms of M, connecting the amount owed at the beginning of the nth year, x_n, with the amount owed at the beginning of the next year, x_{n+1}.

   (iii) Hence find the value of M by generating the sequence of 25 terms.

   (iv) Unfortunately, after 13 years of marriage Mr and Mrs Bifurcation split up. How much of the original mortgage is still outstanding?

3. Each year a survey is made of the number of seals in the Orkney Islands. During the year the number of seals born is approximately 40% of the population at the beginning of the year and the death rate is approximately 15% of the population. At the beginning of the 1990 survey the population of seals numbered exactly 5000.

   (i) Formulate a recurrence relation describing the number of seals at the beginning of a year.

   (ii) Estimate the population in January 1995.

   (iii) In what year will the population of seals reach 20 000?

4. A colony of birds currently has a stable population. Prior to this situation the population increased from an initially low level. When the population was 10 000 the proportionate birth rate was 50% per year and the proportionate death rate was 10% per year. When the population was 20 000 the proportionate birth rate was 30% and the proportionate death rate was 20%.

   A model of the population is based on the following assumptions:

   (i) there is no migration and no exploitation (such as shooting);

   (ii) the proportionate birth rate is a decreasing linear function of population;

   (iii) the proportionate death rate is an increasing linear function of population.

   Show that a model based on these assumptions and the above data predicts that the population grows according to the logistic model and find the stable population size.

   Shooting of the birds is now allowed. This culling is carefully monitored so that the proportionate growth decreases at a further 20% per year. Derive the recurrence relation which models the exploited population. Find the equilibrium population. By solving the recurrence relation using a calculator or computer software package (such as a spreadsheet), predict how many years of hunting will take place before the population is first within 1% of the equilibrium population.

5. The population of fish in a large lake has been stable for some time. Prior to this situation the population was decreasing from an initially relatively high level. When the population was 4000 the proportionate birth rate was 10% and the proportionate death rate was 70%. When the population was 3000 the proportionate birth rate was 30% and the proportionate death rate was 60%.

A model of the population is based on the following assumptions:

(i) there is no exploitation and no restocking;

(ii) the proportionate birth rate is a decreasing linear function of the population;

(iii) the proportionate death rate is an increasing linear function of population.

Show that the model based on these assumptions and the above data predicts that the population falls according to the logistic model; find the equilibrium population size.

Restocking of the lake now takes place at a rate of 20% of the population per year. Derive the recurrence relation which models this situation and find the equilibrium population. How many years will it be before the population is within 5% of the new equilibrium population?

Sketch the graph of the population as a function of time.

6. Use the recurrence relation

$$x_{n+1} = 5x_n - 2$$

to generate x_5 for each of the initial conditions below. Comment on your answers.

(i) $x_0 = 1.0$ (ii) $x_0 = 1.1$
(iii) $x_0 = 0.9$ (iv) $x_0 = 0.5$
(v) $x_0 = 0.51$ (vi) $x_0 = 0.49$

7. Each of the recurrence relations below represents a possible rearrangement of the equation

$$x^3 - 4x^2 + x - 1 = 0$$

By generating the first few terms for each case, with $x_0 = 1$, decide which arrangement is best suited to the solution of the original equation.

(i) $x_{n+1} = 4 - \dfrac{x_n - 1}{x_n^2}$ (ii) $x_{n+1} = 1 + 4x_n^2 - x_n^3$

(iii) $x_{n+1} = (4x_n^2 - x_n + 1)^{\frac{1}{3}}$ (iv) $x_{n+1} = \sqrt{\dfrac{4x_n^2 - x_n + 1}{x_n}}$

4.2 Some terminology

The general form of a first order difference equation model for a discrete dynamical system is

$$x_{n+1} = F(x_n) \tag{4.1}$$

In such an equation the new value of x is determined completely by the previous value. In higher order difference equations we would require information about several previous values to determine the current value. For example,

$$x_{n+1} = x_n - x_{n-1}$$

is a second order difference equation because we require the two previous values x_n, x_{n-1} to find the next value x_{n+1}.

The function F in equation (4.1) is called the **map** or **iteration function** of the system. We solve equation (4.1) by the process of iteration (see Chapter 1, page 18, for a review of this process). For the logistic population model $F(x) = kx(1 - x)$ and for a model of compound interest $F(x) = \left(1 + \frac{r}{100}\right) x$ where r is the percentage interest rate.

The difference equation (4.1) is **linear** if $F(x)$ is a linear function of x, otherwise it is **non-linear**. For the logistic model, $F(x) = kx(1 - x)$ is non-linear and the compound interest model is linear.

The sequence of values generated by difference equation (4.1) with initial value x_0 is called **the orbit of x_0 under mapping F** (or simply the orbit). The initial value x_0 is called **the seed of the orbit.**

The iterative process of generating an orbit involves evaluating the function $F(x)$ repeatedly. The notation for this process is written in the following way:

the second iterate of $F(x)$ is $F(F(x)) = F^2(x)$
the third iterate of $F(x)$ is $F(F(F(x))) = F^3(x)$
the fourth iterate of $F(x)$ is $F(F(F(F(x)))) = F^4(x)$
the nth iterate of $F(x)$ is $F^n(x)$.

It is very important not to interpret $F^n(x)$ as $F(x)$ raised to the power n. The following example clearly illustrates this point.

Example 2

For the map $F(x) = 2x + 1$ write down as a polynomial $F^2(x)$, $F^3(x)$, $F^4(x)$ and $F^n(x)$. Show that this is not equal to $(F(x))^n$.

Solution

If $F(x) = 2x + 1$,

$$F^2(x) = F(F(x)) = F(x) + 1 = 2(2x + 1) + 1 = 4x + 3$$

$$F^3(x) = F(F^2(x)) = 2F^2(x) + 1 = 2(4x + 3) + 1 = 8x + 7$$

$$F^4(x) = 2(8x + 7) + 1 = 16x + 15$$

Hence

$$F^n(x) = 2^n x + 2^n - 1$$

Now

$$(F(x))^2 = (2x + 1)^2 = 4x^2 + 4x + 1 \neq F^2(x)$$

$$(F(x))^3 = (2x + 1)^3 = 8x^3 + 12x^2 + 6x + 1 \neq F^3(x)$$

and so on. So in general

$$(F(x))^n \neq F^n(x)$$

TUTORIAL PROBLEM 4.3

Consider the difference equation

$$x_{n+1} = \sqrt{x_n}$$

(i) Write down the map for this equation.

(ii) Find the first eight terms of the orbit of the map with seed 256.

(iii) For this map, find the expressions $F^2(x)$, $F^3(x)$ and $F^4(x)$. Show that in each case $F^n(x) \neq (F(x))^n$.

The first part of this tutorial problem introduces another important property of difference equations. The orbit appears to be converging to the value $x_n = 1$. Note that if we had started with $x_0 = 1$ then $x_1 = \sqrt{x_0} = 1$; $x_2 = \sqrt{x_1} = 1$ and so on. The number 1 is said to be a fixed point of the system $x_{n+1} = \sqrt{x_n}$. For this system $x = 0$ is also a fixed point.

In general, the orbit of a dynamical system

$$x_{n+1} = F(x_n)$$

is a **fixed point** x_f if $x_f = F(x_f)$. Since the fixed point satisfies $F(x_0) = x_0$ then

$$F^2(x_0) = F(F(x_0)) = F(x_0) = x_0$$

and in general $F^n(x_0) = x_0$. Hence the orbit of a fixed point is the sequence $x_0, x_0, x_0, \ldots$ This is the same concept as introduced in Chapter 2 for continuous systems. At a fixed point the dynamical system is fixed, it never moves. To find the fixed points of a system we solve the equation $x = F(x)$.

Example 3

Find the fixed points of the system

$$x_{n+1} = 2(x_n - 2)^2 - 4$$

Solution

The map is $F(x) = 2(x - 2)^2 - 4$. The fixed points are given as solutions to the equation

$$x = F(x)$$

$$\Rightarrow \quad x = 2(x - 2)^2 - 4$$

$$\Rightarrow \quad 2x^2 - 9x + 4 = 0$$

$$\Rightarrow \quad (2x - 1)(x - 4) = 0$$

$$\Rightarrow \quad x = \tfrac{1}{2} \text{ or } x = 4$$

The fixed points of this system are $x_f = \frac{1}{2}$ and $x_f = 4$.

Consider the system

$$x_{n+1} = a - x_n$$

with seed x_0. Then $x_1 = a - x_0$ and $x_2 = a - x_1 = a - (a - x_0) = x_0$. After two iterations the system returns to x_0. If we carry on the iterations then $x_3 = a - x_2 = a - x_0$ and $x_4 = a - x_3 = x_0$ again. In fact after each pair of iterations the system returns to x_0. The orbit consists of the sequence

$$x_0, \quad a - x_0, \quad x_0, \quad a - x_0, \quad x_0, \quad \ldots$$

for any seed x_0.

Such an orbit is called a **periodic orbit** or **cycle** with period 2. The points x_0 and $a - x_0$ are **periodic points**.

In general the point x_0 is said to be a periodic point of fundamental period k if $F^k(x_0) = x_0$. The orbit is called a **period *k* orbit** or **a *k* cycle** if the orbit returns to x_0 in no fewer than k iterations. A fixed point is a periodic point of period 1 and its cycle is a 1 cycle.

Example 4

Show that -1 lies on a cycle of period 2 for the system

$$x_{n+1} = x_n^2 - 1$$

Find the 2-cycle.

Solution

The orbit with seed $x_0 = -1$ is

$$x_1 = x_0^2 - 1 = 0$$

$$x_2 = x_1^2 - 1 = 0 - 1 = -1$$

and $x_2 = x_0$. The sequence of iterations is

$$-1, \quad 0, \quad -1, \quad 0, \quad -1, \quad 0, \quad \ldots$$

(This is the 2-cycle and -1 and 0 are periodic points of fundamental period 2. Note that they also have period 4, 6, 8, ...; it is the smallest period that is the fundamental period.)

If we had no idea of the values of the periodic points then we could solve this problem by finding values of x_0 for which $F^2(x_0) = x_0$. Since

$$F(x_0) = x_0^2 - 1$$

$$F(F(x_0)) = (F(x_0))^2 - 1$$

$$= (x_0^2 - 1)^2 - 1$$

$$= x_0^4 - 2x_0^2$$

The equation $F^2(x_0) = x_0$ leads to the fourth order equation

$$x_0^4 - 2x_0^2 - x_0 = 0$$

We can see that $x_0 = 0$ is one solution and $x_0 = 1$ is a second. Factorizing the left hand side gives

$$x_0(x_0 + 1)(x_0^2 - x_0 - 1) = 0$$

The roots of $x_0^2 - x_0 - 1 = 0$ are $x_0 = \dfrac{1 \pm \sqrt{5}}{2}$ and these two values of x_0 are the fixed points (i.e. points of period 1) for the system. So this method of approach gives all the periodic points up to period 2.

In general it is very difficult to find all the periodic points in this way. Figure 4.1 shows the search for periodic points of period higher than 2. The graphs show the functions $y = x$, $y = F^3(x)$ and $y = F^4(x)$.

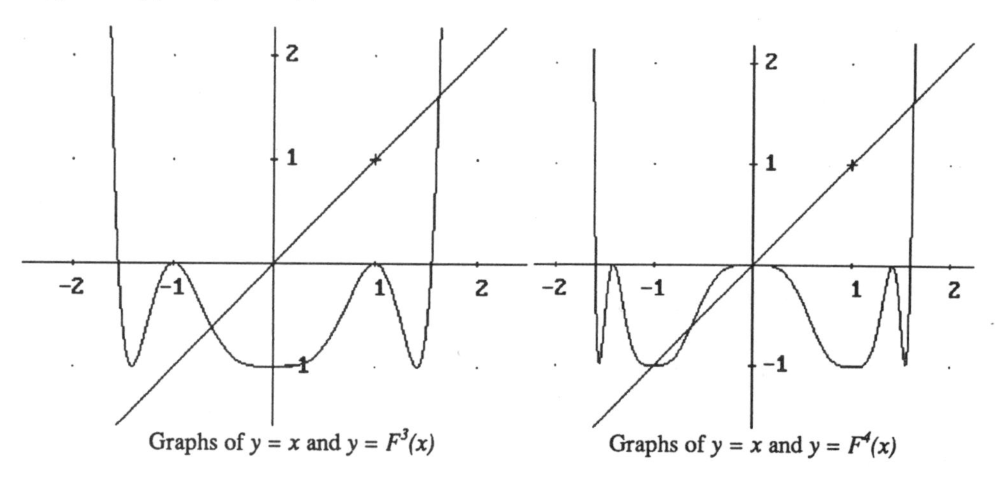

Fig 4.1 Algebraic expressions and associated graphs for period 3 and period 4 orbits

We see that $F^3(x)$ is a polynomial of degree 8 and $F^4(x)$ is degree 16 and the polynomial equations $F^3(x) = x$ and $F^4(x) = x$ of such high degree are usually impossible to solve exactly. The intersections of the graphs of $y = x$ and $y = F^3(x)$ show that the only periodic points are the fixed points $\frac{1 \pm \sqrt{5}}{2}$, and those of the graphs of $y = x$ and $y = F^4(x)$ show that the periodic points are the fixed points and the period-2 points are 0, −1. From the definition of periodic points the fixed periods are periodic of period k for all positive integers and the period-2 points have period $2k$ for all positive integers k. So we would expect that 0, −1 are period-4 points satisfying $F^4(x) = x$.

This initial section of terminology has introduced some important ideas. The questions we need to explore are:

(i) How does the orbit depend on the seed x_0?

(ii) How does the orbit depend on the map $F(x)$?

TUTORIAL PROBLEM 4.4

Show that 1 lies on a cycle of period 3 for the system

$$x_{n+1} = -3x_n^2 + \frac{5}{2}x_n + \frac{1}{2}$$

Write down the 3-cycle.

Find the polynomial equation for the 3-cycle

$$F^3(x) - x = 0$$

Exercises 4.2

1. For each of the following systems identify the map $F(x)$ and find the expressions for $F^2(x)$. For (i) and (ii) find expressions for $F^3(x)$. Find all the fixed points (if any) of the systems.

(i) $x_{n+1} = x_n + 3$ (ii) $x_{n+1} = x_n^2 - 2x_n + 2$

(iii) $x_{n+1} = x_n^3 - 3x_n$ (iv) $x_{n+1} = \cos x_n$

(v) $x_{n+1} = |x_n|$

2. Compute the first five points of the orbit for each of the following systems with given seed x_0:

(i) $x_{n+1} = x_n^2$ $\qquad x_0 = 0.5$

(ii) $x_{n+1} = 1 + x_n^2$ $\qquad x_0 = 0$

(iii) $x_{n+1} = \sin x_n$ $\qquad x_0 = 1$

(iv) $x_{n+1} = \begin{cases} 2x_n & 0 \le x_n < \frac{1}{2} \\ 2x_n - 1 & \frac{1}{2} \le x_n < 1 \end{cases}$ $\qquad x_0 = \frac{1}{4}$

3. Show that $x_0 = \sqrt{2}$ lies on a period-2 orbit of the system

$$x_{n+1} = x_n^3 - 3x_n$$

Find $F^2(x)$ and by drawing appropriate graphs show that there are six periodic points of fundamental period-2.

4. Consider the system

$$x_{n+1} = 0.5x_n + 1$$

(i) Find the fixed point of this system.

(ii) Compute $F^2(x)$, $F^3(x)$, $F^4(x)$, $F^n(x)$.

(iii) Deduce that there are no periodic points other than the fixed points.

5. Consider the system

$$x_{n+1} = x_n^2 - x_n - 2$$

(i) Find the fixed points of this system.

(ii) Show that $\sqrt{2}$ and $-\sqrt{2}$ are periodic points of period-2.

(iii) Show graphically that there are six periodic points of period-3. Find their values correct to 2 decimal places.

6. This problem investigates the map defined by the system

$$x_{n+1} = \begin{cases} 2x_n & 0 \le x_n \le \frac{1}{2} \\ 2(1-x_n) & \frac{1}{2} \le x_n \le 1 \end{cases}$$

This is called the **tent map** and denoted by $T(x)$.

(i) Find the fixed points of the tent map.

(ii) Find the formulas for $T^2(x)$ and $T^3(x)$.

(iii) Sketch the graphs of $T(x)$, $T^2(x)$ and $T^3(x)$. Are there any period-2 or period-3 orbits.

(iv) Investigate $T^n(x)$ for different values of n. Can you write down a general formula for $T^n(x)$.

7. This problem investigates the map defined by the system

$$x_{n+1} = D(x_n) = \begin{cases} 2x_n & 0 \le x_n < \frac{1}{2} \\ 2x_n - 1 & \frac{1}{2} \le x_n < 1 \end{cases}$$

This is called the **doubling map** since all points within the domain [0,1) are doubled.

(i) Show that 0 is the only fixed point.

(ii) Show that $\frac{1}{3}$ is a periodic point of period-2. Write down the 2-cycle.

(iii) Show that $\frac{1}{5}$ is a periodic point of period-4. Write down the 4-cycle.

(iv) Show that $\frac{1}{9}$ is a periodic point of period-6. Write down the 6-cycle.

(v) Find formulas for $D^2(x)$, $D^3(x)$, $D^4(x)$ and sketch their graphs. Are there any 3-cycles?

(vi) Investigate $D^n(x)$ for different values of n. Can you write down a general formula for $D^n(x)$?

4.3 Linear discrete systems

A study of linear discrete systems will introduce many of the basic concepts for the more mathematically rich non-linear systems. The general form of a linear difference equation is

$$x_{n+1} = a + bx_n$$

where a and b are constants. The **linear map is $F(x) = a + bx$.**

TUTORIAL PROBLEM 4.5

Consider the linear map

$$x_{n+1} = bx_n$$

(i) Find the solution for the seed x_0.

(ii) Show that if $b \neq 1$ then 0 is the only fixed point and that there are no periodic cycles. If $b = 1$ show that x_0 is a fixed point for all choices of x_0.

(iii) Show that if $|b| > 1$ the fixed point is unstable and if $|b| < 1$ the fixed point is stable.

Consider the general linear map $F(x) = a + bx$. The fixed points of this map are given as solutions to $x = a + bx$. There is one fixed point for each pair of values of a and b:

$$x_f = \frac{a}{1-b} \qquad b \neq 1$$

If $b = 1$ then the graphs of $y = x$ and $y = F(x)$ are parallel and there are no fixed points if $a \neq 0$.

The type of orbit for linear map depends on the size of b as shown in the graphs of Figure 4.2.

For $b > 0$ the graphs are staircase diagrams and for $b < 0$ the graphs are cobweb diagrams. For $|b| < 1$ the fixed point is stable since the orbit converges to x_f. This is an example of an **attractor**. For $|b| > 1$ the fixed point is unstable since the orbit diverges. This is an example of **a repellor**.

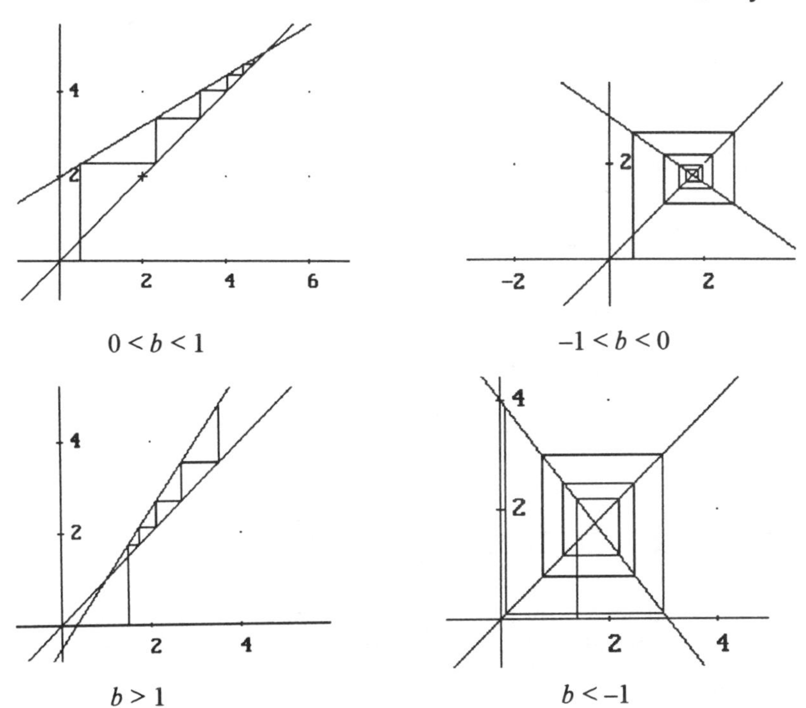

Fig 4.2 Graphical analysis of the linear map $x_{n+1} = a + bx_n$

Figure 4.3 shows the graphical analysis for $b = -1$, i.e. the system $x_{n+1} = a - x_n$, for which every orbit is a 2-cycle.

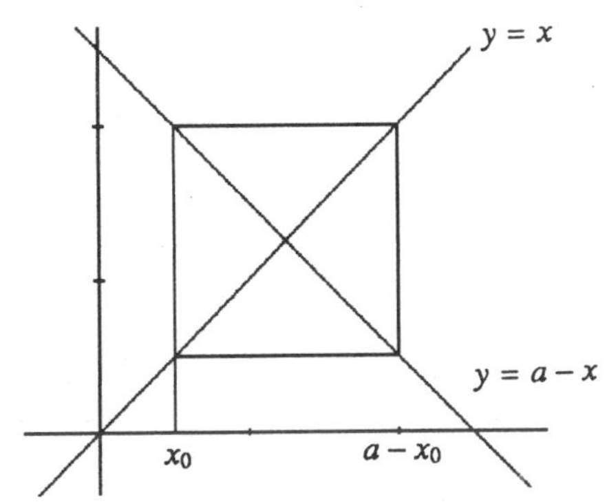

Fig 4.3 A 2-cycle for the map $F(x) = a - x$

Exercises 4.3

1. For the linear map $F(x) = a + bx$ find $F^2(x)$ and $F^3(x)$.

   (i) By solving $F^2(x) = x$ show that $b = -1$ gives period-2 cycles for any choice of x_0.

   (ii) Show that there are no period-3 cycles.

2. For the linear map $F(x) = a + bx$ find an expression for $F^n(x)$. Sketch the graphs of $F(x)$ and $F^n(x)$ for $|b| < 1$ and $|b| > 1$.

3. Consider the system

$$x_{n+1} = F(x_n) = |x_n|$$

(i) Find expressions for $F^2(x)$ and $F^3(x)$.

(ii) Find any fixed points for the system.

(iii) Are there any periodic orbits?

4.4 Non-linear discrete systems

The most interesting and often surprising behaviour occurs for non-linear mapping functions. To give a flavour of what lies ahead consider the quadratic function $F(x) = x^2 - 2$, so that

$$x_{n+1} = x_n^2 - 2$$

First of all the orbit with seed $x_0 = 0$ is very simple

0, –2, 2, 2, 2, 2

The point $x = 2$ is one of the fixed points of this system ($x = -1$ is the other fixed point). However, consider the orbits with seeds x_0 close to 0. Table 4.1 shows the orbits for $x_0 = 0.01$ and $x_0 = 0.1$. After a few iterations the orbits seem to wander randomly all over the place. However, notice that the orbits are contained within the interval (–2,2).

Seed 0.01		Seed 0.1	
0.01	0.7929	0.1	1.1538
–1.9999	–1.3712	–1.99	–0.6688
1.9996	–0.1197	1.9601	–1.5527
1.9984	–1.9857	1.8420	0.4109
1.9936	1.9429	1.3929	–1.8311
1.9745	1.7748	–0.0597	1.3531
1.8985	1.1498	–1.9964	–0.1692
1.6042	–0.6780	1.9857	–1.9714
0.5734	–1.5403	1.9432	1.8864
–1.6712	0.3724	1.7759	1.5584

Table 4.1 Orbits of the map $F(x) = x^2 - 2$

The table of values shows that even for such a simple looking system $x_{n+1} = x_n^2 - 2$ the behaviour of the orbits is somewhat strange. This is in fact our first view of chaotic behaviour of a system.

The fixed points of the system are $x_f = -1$ and $x_f = 2$. Figure 4.4 shows the graphical analysis for three initial values $x_0 = -2.5$, $x_0 = -1.5$ and $x_0 = 2.5$.

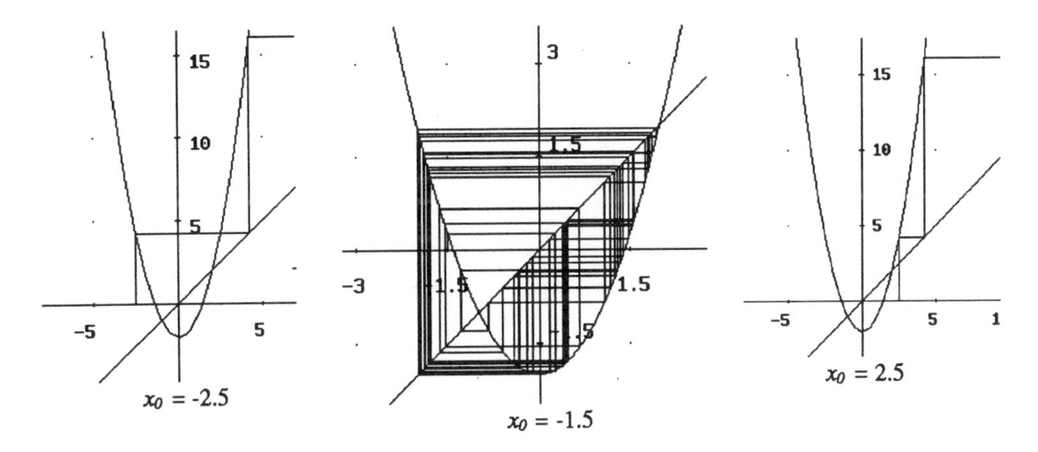

Fig 4.4 Graphical behaviour of $x_{n+1} = x_n^2 - 2$

For initial values outside the interval (–2,2) the fixed points appear to be unstable. However for the initial value within the interval (–2,2) the orbit is very complicated. The orbit remains within this interval but does not converge to either of the fixed points. This is a second glimpse of chaotic behaviour.

The graphical analysis helps us to understand the behaviour of many systems but is not a rigorous method of approach. We now develop an algebraic approach which uses the properties of linear systems of the previous section.

Suppose that $x_{n+1} = F(x_n)$ is a non-linear system, so that $F(x)$ is a non-linear function of x. Let x_f be a fixed point of the system.

Consider a Taylor polynomial expansion of $F(x)$ about the point $x = x_f$. We have

$$F(x) \approx F(x_f) + (x - x_f)F'(x_f) + \frac{(x - x_f)^2}{2}F''(x_f) + \ldots$$

Since x_f is a fixed point, $F(x_f) = x_f$, so we can write $F(x)$ as

$$F(x) \approx x_f + (x - x_f)F'(x_f) + 0(x - x_f)^2$$

$$= x_f(1 - F'(x_f)) + xF'(x_f) + 0(x - x_f)^2$$

For a neighbourhood close to x_f, the linearized approximation to $F(x)$ is

$$F(x) = x_f\,(1 - F'(x_f)) + xF'(x_f) \qquad F'(x_f) \neq 0$$

Comparing the linearized system

$$x_{n+1} = x_f(1 - F'(x_f)) + x_n F'(x_f)$$

with $x_{n+1} = a + bx_n$ we have

$$a = x_f(1 - F'(x_f)) \quad \text{and} \quad b = F'(x_f)$$

We conclude that the fixed point x_f is stable for $|F'(x_f)| < 1$ and is unstable for $|F'(x_f)| > 1$. If $|F'(x_f)| = 1$ then the linearization does not give a reliable method of approach. A fixed point for which $F'(x_f) \neq \pm 1$ is said to be **simple**; if $F'(x_f) = \pm 1$ the fixed point is **not simple**.

So there are two markedly different types of fixed points.

A fixed point that attracts all points sufficiently close to the fixed point is called an **attractor**.

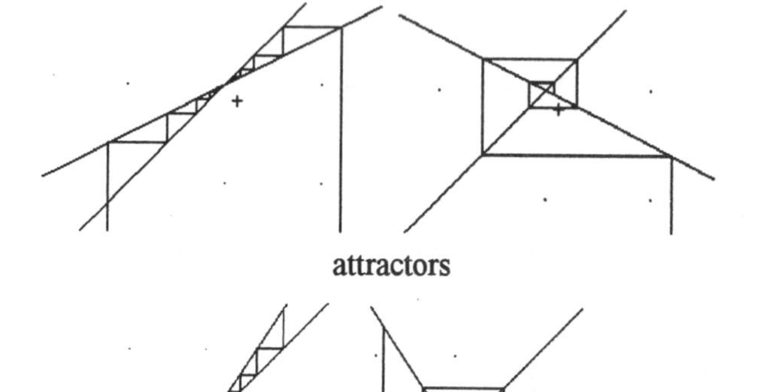

attractors

A fixed point that repels all points sufficiently close to the fixed point is called a **repellor**.

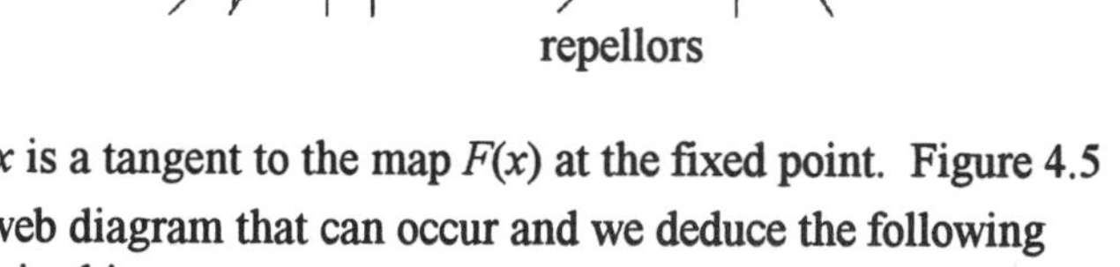

repellors

For $F'(x_f) = 1$ the line $y = x$ is a tangent to the map $F(x)$ at the fixed point. Figure 4.5 shows the four types of cobweb diagram that can occur and we deduce the following criteria for the fixed point x_f in this case.

If $f'(x_f) = 1$ and $f''(x_f) > 0$ then the sequence converges from below $x_0 < x_f$ and diverges from above $x_0 > x_f$. We say that x_f is **semi-stable from below**.

If $f'(x_f) = 1$ and $f''(x_f) < 0$ then x_f is **semi-stable from above**.

For the case of a point of inflexion $f'(x_f) = 1, f''(x_f) = 0$; if $f'''(x_f) > 0$ then x_f is unstable and if $f'''(x_f) < 0$ then x_f is stable.

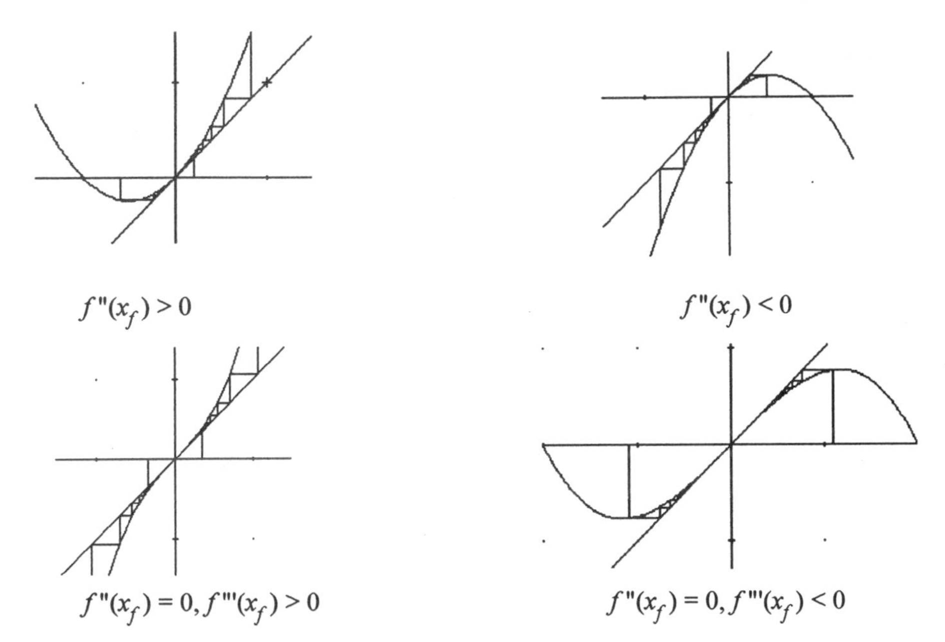

Fig 4.5 The cobweb diagrams for $F(x)$ near the fixed point x_f where $F'(x_f) = 1$

For $F'(x_f) = -1$ the possible options on stability are intuitively obvious but a careful mathematical analysis is necessary to develop a criteria for stability. Figure 4.6 shows that the fixed point is either stable or unstable. The criterion, which we will not derive here, is that x_f is stable if $-2F'''(x_f) - 3(F''(x_f))^2 < 0$.

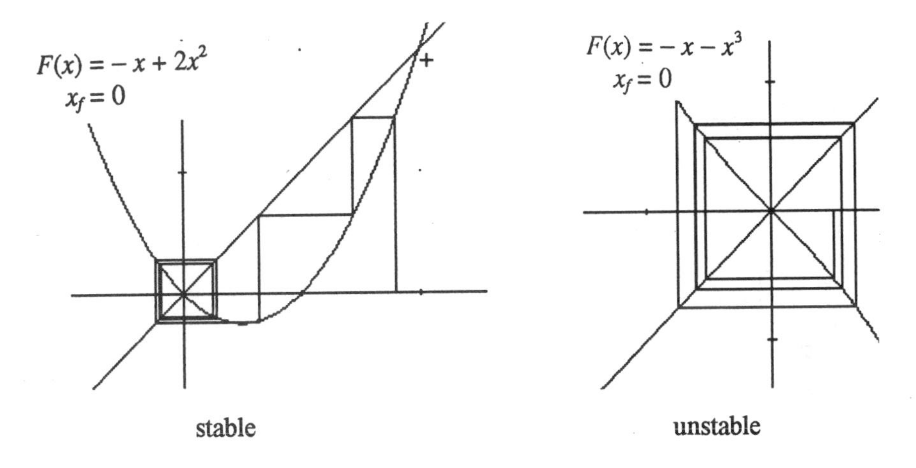

Fig 4.6 The fixed point x_f is stable or unstable if $F'(x_f) = -1$

Now for many systems we can account for the behaviour of the orbits of all points and classify the fixed points. When we can do this then we say that we have carried out a complete **orbit analysis**. As we have seen for the simple looking quadratic system $x_{n+1} = x_n^2 - 2$, an orbit analysis will produce some behaviour which we do not completely understand. In the next section we explore the quadratic system in some detail.

Example 5

Give an orbit analysis for the system

$$x_{n+1} = x_n^3$$

Solution

The fixed points are given as solutions to the equation

$$x^3 = x$$

There are three fixed points, 0, 1 and −1. The stability of the fixed points are defined by the size of $F'(x_f)$ where $F(x) = x^3$.

At $x = -1$ $\quad F'(-1) = 3x^2|_{-1} = 3 > 1$

At $x = 0$ $\quad F'(0) = 3x^2|_0 = 0 < 1$

At $x = 1$ $\quad F'(1) = 3x^2|_1 = 3 > 1$

The fixed point at $x = 0$ is stable; the fixed points at $x = -1$ and $x = 1$ are unstable. For $-1 < x_0 < 1$ the orbit converges to $x = 0$. On the other hand if $|x_0| > 1$ the orbit tends to $\pm\infty$. Figure 4.7 shows the orbit analysis.

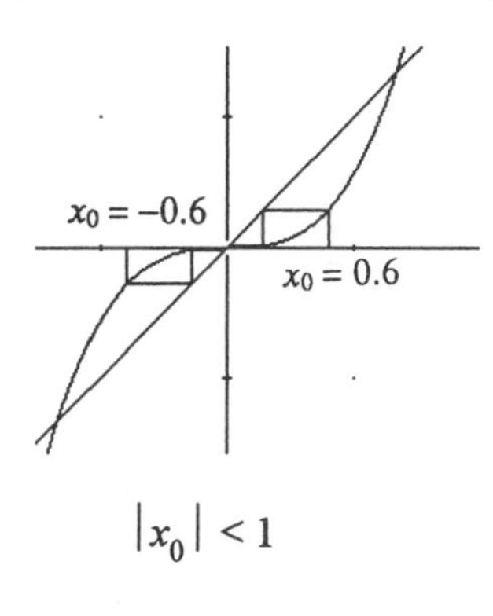

$|x_0| < 1$

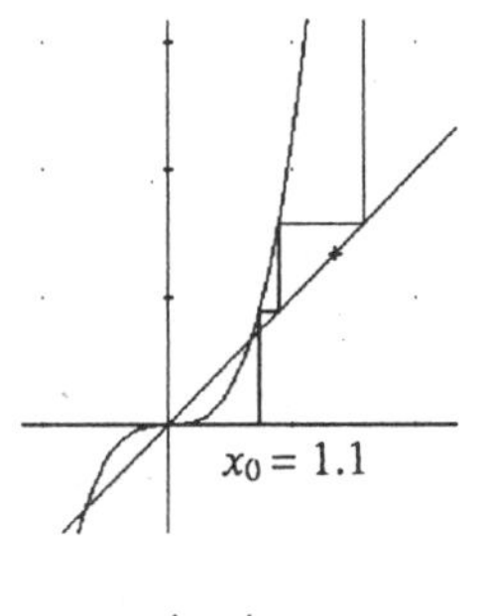

$|x_0| > 1$

Fig 4.7 Orbit analysis for $x_{n+1} = x_n^3$

TUTORIAL PROBLEM 4.6

Give an orbit analysis for the system

$$x_{n+1} = \frac{1}{a + x_n} \qquad x_0 \geq 0$$

where a is a positive constant.

This example and tutorial problem suggest that for some initial values the sequence converges to a stable first point and for others the sequence diverges to $\pm\infty$. But there are other types of points.

A point x_0 is called an **eventually fixed point** if x_0 itself is not fixed but some point on the orbit of x_0 is fixed. For example, the system $x_{n+1} = x_n^2$ has fixed points 0 and 1. For initial values $|x_0| > 1$ the orbit converges to the fixed point $x_f = 0$. For $|x_0| > 1$ the orbit diverges. For $x_0 = -1$ the orbit is -1, 1, 1, 1, ... So $x_0 = -1$ is eventually fixed.

A point x_0 is called an **eventually periodic point** if x_0 itself is not periodic but some point on the orbit of x_0 is periodic. For example, the system $x_{n+1} = x_n^2 - 1$ has periodic orbit 0, -1, 0, -1, ... so that 0 and -1 are period points of period 2. For $x_0 = \sqrt{2}$ the orbit is $\sqrt{2}$, 1, 0, 1, 0, ... So $x_0 = \sqrt{2}$ is eventually periodic.

TUTORIAL PROBLEM 4.7

Consider the map $F(x) = -\frac{3}{2}x^2 + \frac{5}{2}x + 1$.

(i) Show that 0 lies on a 3-cycle and find the orbit of fundamental period 3.

(ii) Show that $x_0 = \frac{1}{3}$, $x_0 = \frac{4}{3}$, $x_0 = \frac{2}{3}$ are eventually periodic points.

(iii) Suggest a method for finding eventually periodic and eventually fixed points.

In the same way as fixed points can be attracting or repelling, periodic cycles may also be classified in this way. For example, Fig 4.8 shows the cobweb diagram for the orbit of $x_0 = 0.5$ for the system $x_{n+1} = x_n^2 - 1$. We see that the orbit is attracted towards the 2-cycle 0, -1, 0, -1 ... This is an example of an attracting periodic cycle.

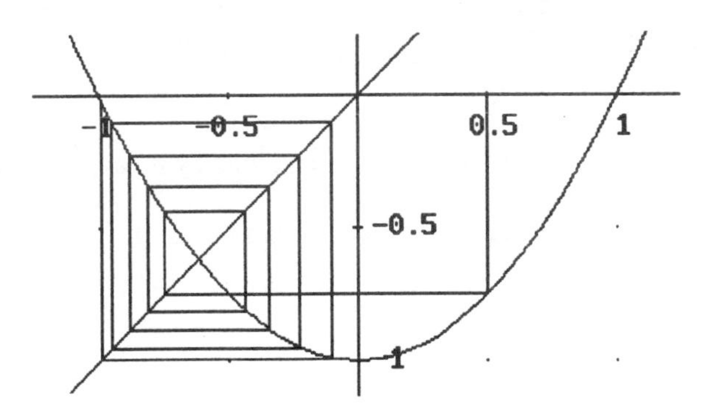

Fig 4.8 An attracting 2-cycle for $x_{n+1} = x_n^2 - 1$

To examine this 2-cycle consider the graph of $F^2(x)$ shown in Fig 4.9. Since $F(x) = x^2 - 1$ we have

$$F^2(x) = (x^2 - 1)^2 - 1 = x^4 - 2x^2$$

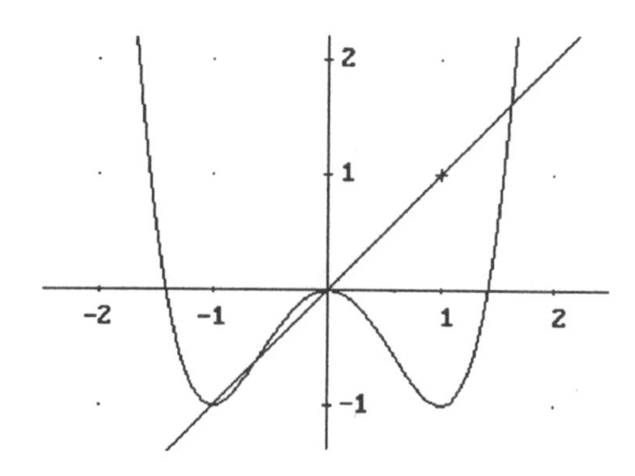

Fig 4.9 The graph of $F^2(x) = x^4 - 2x^2$

The graph of $F^2(x)$ shows four fixed points, $\frac{1}{2}(1 \pm \sqrt{5})$, 0 and −1. These are solutions of the equation $F^2(x) = x$. The two points $\frac{1}{2}(1 \pm \sqrt{5})$ are the fixed points of the function $F(x)$. The extra two fixed points 0, −1 are the period-2 points.

Since $\left|\frac{dF}{dx}\right| = |2x| > 1$ for each fixed point then $x_f = \frac{1}{2}(1 \pm \sqrt{5})$ are repelling fixed points.

Consider the value of $\frac{dF^2}{dx}$ at each of the period-2 points. Since $F^2(x) = x^4 - 2x^2$ then $\frac{dF^2}{dx} = 4x^3 - 4x$. At $x = 0$, $\frac{dF^2}{dx} = 0$ and at $x = -1$, $\frac{dF^2}{dx} = 0$.

Hence the periodic points 0 and −1 are attracting fixed points of the map $F^2(x)$. So under iteration of the map $F(x)$ the sequence cycles back and forth as it converges to the period 2-cycle. The cobweb diagram in Fig 4.8 shows this behaviour.

This example allows us to generalize the ideas of attracting and repelling cycles. A periodic point of fundamental period k is attracting (repelling) if it is an attracting (repelling) fixed point of the map $F^k(x)$.

The periodic point is classified as attracting or repelling by investigating the size of $\frac{dF^k}{dx}$. If $\left|\frac{dF^k}{dx}\right| < 1$ at the periodic point then it is attracting. If $\left|\frac{dF^k}{dx}\right| > 1$ at the periodic point then it is repelling.

TUTORIAL PROBLEM 4.8

(i) Suppose that $x_0, x_1, \ldots$ lie on a cycle of period-2 for the map $F(x)$ so that $x_i = F^2(x_i)$ for each i. Show that

$$\frac{dF^2}{dx}(x_0) = \frac{dF}{dx}(x_1)\frac{dF}{dx}(x_0)$$

(ii) For the map $F(x) = x^2 - 1$ there is a period-2 cycle 0, −1. Confirm that this cycle is attracting by evaluating the right hand side of the expression in (i).

(iii) Suppose that x_0, x_1, x_2 lie on a cycle of period-3 for the map $F(x)$. Show that

$$\frac{dF^3}{dx}(x_0) = \frac{dF}{dx}(x_2)\frac{dF}{dx}(x_1)\frac{dF}{dx}(x_0)$$

(iv) For the map $F(x) = -\frac{3}{2}x^2 + \frac{5}{2}x + 1$ there is a period-3 cycle 0, 1, 2. Show that this cycle is repelling. Verify this result by drawing a cobweb diagram for $x_0 = 0.5$.

Example 6

Consider the map

$$F(x) = ax - x^2$$

(i) Find and classify the fixed points.
(ii) Find and classify any period-2 cycles.

Solution

(i) The fixed points are solutions of $F(x) = x$, i.e.

$$ax - x^2 = x$$

solving for x, $x_f = 0$ and $x_f = a - 1$.

To classify the fixed points we investigate $\frac{dF}{dx}$.

For $F(x) = ax - x^2$, $\frac{dF}{dx} = a - 2x$

$$x_f = 0: \qquad \frac{\mathrm{d}F}{\mathrm{d}x} = a$$

The fixed point $x_f = 0$ is stable for $|a| < 1$ and unstable for $|a| > 1$. (For $a = 1$ see Problem 1 (iii).)

$$x_f = a - 1: \qquad \frac{\mathrm{d}F}{\mathrm{d}x} = 2 - a$$

The fixed point $x_f = a - 1$ is stable for $1 < a < 3$ and unstable otherwise.

(ii) To find any period-2 cycles we solve $F(F(x)) = x$. For $F(x) = ax - x^2$

$$F(F(x)) = -x^4 + 2x^3 a - (a^2 + a)x^2 + a^2 x$$

The equation $F(F(x)) = x$ is a quartic polynomial in x which is not easy to solve. However we know that the fixed points will be period-k cycles for any k because they are period-1 cycles. In particular they are period-2 cycles. Hence $F(x) - x$ will be a factor of $F(F(x)) - x$ and $F(F(x)) - x$ divided by $F(x) - x$ will give a quadratic polynomial.

$$\frac{F(F(x)) - x}{F(x) - x} = \frac{-x^4 + 2x^3 a - (a^2 + a)x + (a^2 - 1)x}{-x^2 + (a-1)x}$$

$$= x^2 - x(a+1) + (a+1)$$

So the cycles of fundamental period 2 will be solutions of

$$x^2 - x(a + 1) + (a + 1) = 0$$

Solving for x:

$$x = \frac{(1+a) \pm \sqrt{(a+1)^2 - 4(a+1)}}{2}$$

$$x = \frac{(1+a) \pm \sqrt{(a+1)}\sqrt{a-3}}{2}$$

Providing $a > 3$ there is a period-2 cycle

$$x_0 = \tfrac{1}{2}\sqrt{a+1}(\sqrt{a+1} + \sqrt{a-3}), \quad x_1 = \tfrac{1}{2}\sqrt{a+1}(\sqrt{a+1} - \sqrt{a-3})$$

To classify the type of orbit we investigate the size of $\frac{\mathrm{d}F}{\mathrm{d}x}(x_0)\cdot\frac{\mathrm{d}F}{\mathrm{d}x}(x_1)$. For $F(x) = ax - x^2$, $\frac{\mathrm{d}F}{\mathrm{d}x} = a - 2x$, so

$$\frac{dF}{dx}(x_0)\cdot\frac{dF}{dx}(x_1) = -a^2 + 2a + 4$$

For an attracting period-2 cycle

$$|-a^2 + 2a + 4| < 1$$

Solving this inequality gives

$$3 < a < 1+\sqrt{6}$$

for an attracting period-2 cycle. If $a > 1+\sqrt{6}$ the period-2 cycle is repelling.

Exercises 4.4

1. Examine, by graphical analysis, the behaviour of orbits for each of the following maps:

(i) $F(x) = -3x + 1$ (ii) $F(x) = x + x^2$

(iii) $F(x) = x - x^2$ (iv) $F(x) = x - x^3$

(v) $F(x) = \sin x$

2. Give a complete orbit analysis for each of the following maps:

(i) $F(x) = \frac{1}{2}x + 1$ (ii) $F(x) = \sqrt{x}$

(iii) $F(x) = \frac{1}{x}$ (iv) $F(x) = e^x$

(v) $F(x) = -x^2$ (vi) $F(x) = 1.36x - x^3$

3. Consider the quadratic system

$$x_{n+1} = x_n^2 - 1.1$$

(i) Find and classify the fixed points of this system.

(ii) Write $F^2(x)$ as a polynomial in x.

(iii) Find the periodic cycle of fundamental period 2.

(iv) Sketch some of the orbits.

4. For each of the following maps find the fixed points and some eventually fixed points:

   (i) $F(x) = x(1 - x)$

   (ii) $F(x) = 3x(1 - x)$

   (iii) $F(x) = x^4 - 4x^2 + 2$

   (iv) $F(x) = |x|$

5. Consider the system

$$x_{n+1} = x_n^3 - 3x_n$$

   (i) Find the fixed points of this system.

   (ii) Find the three periodic orbits of fundamental period 2 for this system.

   (iii) Classify each periodic orbit as attracting or repelling. Verify your results by drawing appropriate cobweb diagrams.

   (iv) Show that the points $x_0 = -\frac{1}{2}(\sqrt{2} \pm \sqrt{6})$ are eventually periodic and in each case draw a cobweb diagram showing the orbit.

6. For what values of the constants a and x_0 does the following sequence tend to a limit?

$$x_0,\quad \sqrt{a + x_0},\quad \sqrt{a + \sqrt{a + x_0}},\quad \sqrt{a + \sqrt{a + \sqrt{a + x_0}}},\ \ldots$$

   [Hint: investigate the system $x_{n+1} = \sqrt{a + x_n}$.]

7. For what values of the positive constant a does the sequence

$$a,\quad a^a,\quad a^{a^a},\ \ldots$$

   tend to a limit?

   [Hint: investigate the system $x_{n+1} = a^{x_n}$.]

8. Consider the map

$$F(x) = ax - x^3$$

(i) Find the fixed points.

(ii) Show that this map has a repelling period-2 cycle given by

$$x = \pm\sqrt{a+1} \qquad (a > -1)$$

(iii) Find and classify any other period 2 cycles.

4.5 Quadratic maps

In this section we concentrate on one particular family of non-linear first order systems, those defined by the quadratic polynomial $F(x) = x^2 - c$. We will see that remarkable changes occur as we change the parameter c. We have met some of the orbits of this system in earlier sections. For instance, in Example 4 we showed that $x_{n+1} = x_n^2 - 1$ has the attracting 2-cycle $-1, 0, -1, 0$. Figure 4.10 (a) shows the cobweb diagram for $x_0 = 0.5$. Table 4.1 (page 122) shows the remarkable change in behaviour of $x_{n+1} = x_n^2 - 2$ for $x_0 = 0.01$ and 0.1 and Fig 4.10 (b) shows the cobweb diagram for $x_0 = -1.5$.

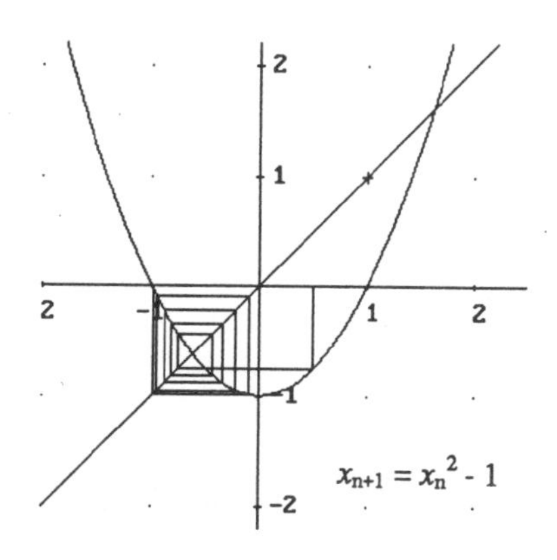

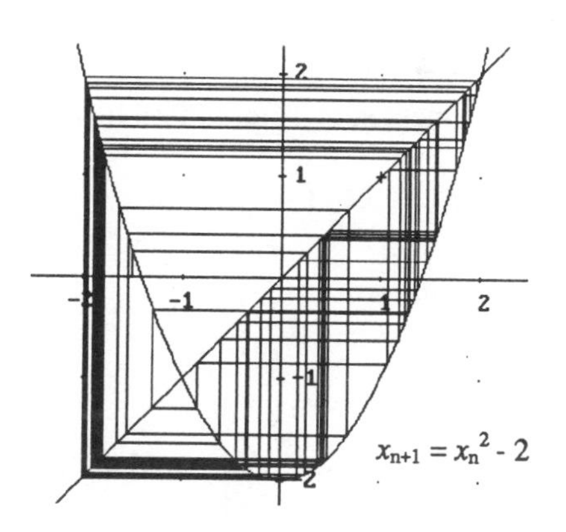

Fig 4.10 (a) Cobweb diagram for $x_{n+1} = x_n^2 - 1$ and $x_0 = 0.5$

(b) Cobweb diagram for $x_{n+1} = x_n^2 - 2$ and $x_0 = -1.5$

The quadratic map $F(x) = x^2 - c$ may appear a special choice. However, Tutorial Problem 4.9 shows that all quadratic first order systems can be reduced to this standard form.

TUTORIAL PROBLEM 4.9

Consider the general non-linear quadratic system

$$x_{n+1} = \alpha + \beta x_n + \gamma x_n^2 \qquad \text{with } \gamma \neq 0 \qquad (4.2)$$

Making the substitution

$$x_n = \lambda + \mu y_n$$

choose values of λ and μ so that the difference equation becomes the standard form

$$y_{n+1} = y_n^2 - c$$

You should have found that, by choosing λ and μ satisfying the two equations $\gamma\mu = 1$ and $\beta + 2\gamma\lambda = 0$,

$$y_{n+1} = y_n^2 - c$$

with $c = \frac{1}{4}\beta^2 - \frac{1}{2}\beta - \alpha\gamma$.

Example 7

Use the method of Tutorial Problem 4.9 to reduce the logistic system

$$x_{n+1} = kx_n(1 - x_n)$$

to standard form.

Solution

Let $x_n = \lambda + \mu y_n$ for some λ and μ. Then $x_{n+1} = \lambda + \mu y_{n+1}$. Substituting for x_{n+1} and x_n we get

$$\lambda + \mu y_{n+1} = k(\lambda + \mu y_n)(1 - \lambda - \mu y_n)$$

$$= k(\lambda(1 - \lambda) + (\mu - 2\lambda\mu)y_n - \mu^2 y_n^2)$$

$$\Rightarrow \qquad y_{n+1} = \frac{(k(1-\lambda)-1)\lambda}{\mu} + (1-2\lambda)y_n - k\mu y_n^2$$

If we choose $\lambda = \frac{1}{2}$ the linear coefficient is zero and choose $k\mu = -1$ and $c = -\lambda(k(1-\lambda) - 1)/\mu$ to give the standard form

$$y_{n+1} = y_n^2 - c$$

With $\lambda = \frac{1}{2}$ and $\mu = -\frac{1}{k}$ we have $c = \frac{k}{4}(k-2)$.

So without any loss in generality we can concentrate our investigation on the quadratic system

$$x_{n+1} = x_n^2 - c \qquad (4.3)$$

TUTORIAL PROBLEM 4.10

(i) Give an orbit analysis for the case $c = 0$.

(ii) Describe the results of iteration with $x_0 = 0$ for each of the cases

(a) $c = -1$, and
(b) $c = 1$.

The results of this tutorial problem are summarized below:

$c = -1$ so that $x_{n+1} = x_n^2 + 1$

There are no fixed points; for all initial values x_0 the iterations diverge to $\pm\infty$.

$c = 0$ so that $x_{n+1} = x_n^2$

There is an attracting fixed point at $x_f = 0$ which attracts all orbits for $-1 < x_0 < 1$. There is a repelling fixed point at $x_f = 1$ and $x_0 = -1$ is an eventually fixed point.

$c = 1$ so that $x_{n+1} = x_n^2 - 1$

There are two repelling fixed points at $\frac{1}{2}(1 \pm \sqrt{5})$ and an attracting 2-cycle, $-1, 0, -1, 0$.

So major changes occur as we change the values of c. We now investigate the critical values of c where these changes occur.

The fixed points

The fixed points of $x_{n+1} = x_n^2 - c$ are given by the equation

$$x = x^2 - c$$

Solving for x,

$$x_f = \frac{1}{2}(1 \pm \sqrt{1+4c}) \tag{4.4}$$

These solutions are real provided $1 + 4c \geq 0$, i.e. $c \geq -\frac{1}{4}$. There are two fixed points and the next step is to classify them as stable or unstable.

To classify these fixed points consider $F'(x) = 2x$.

For $x_f = \frac{1}{2}(1+\sqrt{1+4c})$, $F'(x_f) = 1+\sqrt{1+4c} > 1$ for all $c > -\frac{1}{4}$ and so $x_f = \frac{1}{2}(1+\sqrt{1+4c})$ is unstable.

For $x_f = \frac{1}{2}(1-\sqrt{1+4c})$, $F'(x_f) = 1-\sqrt{1+4c} < 0$. The fixed point is stable for

$$F'(x_f) > -1$$

$$\Rightarrow \quad 1-\sqrt{1+4c} > -1$$

$$\Rightarrow \quad \sqrt{1+4c} < 2$$

$$\Rightarrow \quad c < \frac{3}{4}$$

If $c = \frac{3}{4}$ then $F'(x_f) = -1$ and the fixed point is not simple.

If $c > \frac{3}{4}$ then $F'(x_f) < -1$ and the fixed point is unstable.

Hence for $-\frac{1}{4} < c < \frac{3}{4}$ there are two simple fixed points:

$$x_f^+ = \frac{1}{2}(1+\sqrt{1+4c}) \qquad \text{which is unstable}$$

$$x_f^- = \frac{1}{2}(1-\sqrt{1+4c}) \qquad \text{which is stable}$$

Note that if $x_0 = -x_f^+$ then $x_1 = \frac{1}{2}(1+\sqrt{1+4c}) = x_f^+$. We can deduce that any initial point within the interval

$$-x_f^+ < x_0 < x_f^+ \qquad (4.5)$$

remains within the interval and converges to the fixed point $x_f^- = \frac{1}{2}(1-\sqrt{1+4c})$. The interval defined by equation (4.5) is called **the fundamental interval** and turns out to be very important in our analysis. Any point outside the fundamental interval diverges to ∞. In a sense any point within the fundamental interval is captured by it and cannot escape. Figure 4.11 shows typical orbits for the quadratic map $-\frac{1}{4} < c < \frac{3}{4}$.

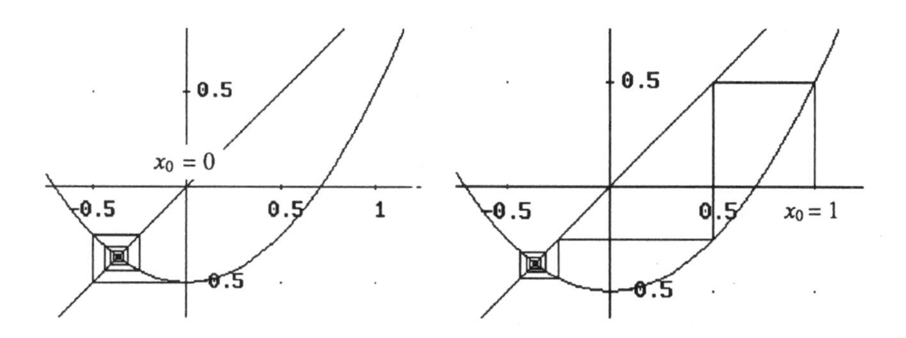

Fig 4.11 The system $x_{n+1} = x_n^2 - c$ for $c = 0.5$ with seed (i) $x_0 = 0$ and (ii) $x_0 = 1$

TUTORIAL PROBLEM 4.11

Investigate the orbits of the system

$$x_{n+1} = x_n^2 - \frac{3}{4}$$

The birth of a period-2 orbit

The next question is what happens when $c = \frac{3}{4}$? We know that x_f^- will remain a fixed point but it is now unstable. So we have two unstable fixed points. We have already seen that for $c = 1$ there is a period-2 cycle: 0, –1, 0, –1, ... To investigate the birth of period-2 cycles consider $F^2(x) = x$.

$$F^2(x) = (x^2 - c)^2 - c = x^4 - 2x^2c + c^2 - c = x$$

i.e. we have to solve the polynomial equation

$$x^4 - 2x^2c - x + c^2 - c = 0 \tag{4.6}$$

The fixed points of $F(x)$ will satisfy this equation because a fixed point is a 1-cycle and is hence a k-cycle for any positive integer k. Factorizing equation (4.6),

$$\underbrace{(x^2 - x - c)}_{\substack{\uparrow \\ \text{equation for} \\ \text{fixed points}}} \quad \underbrace{(x^2 + x - (c-1))}_{\substack{\uparrow \\ \text{gives the} \\ \text{period-2 cycles}}} = 0$$

The period-2 cycles are solutions of

$$x^2 + x - (c - 1) = 0$$

$$\Rightarrow \qquad x = \frac{1}{2}(-1 \pm \sqrt{4c-3})$$

which are real provided $c \geq \frac{3}{4}$. A period-2 cycle is born when $c = \frac{3}{4}$ and the orbit is $\frac{1}{2}(-1-\sqrt{4c-3})$, $\frac{1}{2}(-1+\sqrt{4c-3})$. For example when $c = 1$ the 2-cycle is –1, 0, as before.

The 2-cycle is attracting if $\left|\frac{\mathrm{d}F^2}{\mathrm{d}x}(x)\right| < 1$ at the period points. Since $F^2(x) = x^4 - 2x^2c + c^2 - c$, $\frac{\mathrm{d}}{\mathrm{d}x}F^2(x) = 4x^3 - 4xc = 4x(x^2 - c)$. If x_p^+ and x_p^- are the periodic points then

$$\frac{d}{dx}F^2(x) = 4x_p^+x_p^-$$

$$= 4\left[\frac{1}{2}(-1-\sqrt{4c-3})\right]\left[\frac{1}{2}(-1+\sqrt{4c-3})\right]$$

$$= 1-(4c-3)$$

$$= 4-4c$$

and the 2-cycle is attracting if $|4-4c| < 1$.

$\Rightarrow \quad -1 < 4-4c < 1$

$\Rightarrow \quad \frac{3}{4} < c < \frac{5}{4}$

The period 2-cycle exists for $c < \frac{5}{4}$. Figure 4.12 shows typical orbits for $\frac{3}{4} < c < \frac{5}{4}$. Orbits that start inside remain inside the fundamental interval and are attracted to the period-2 cycle.

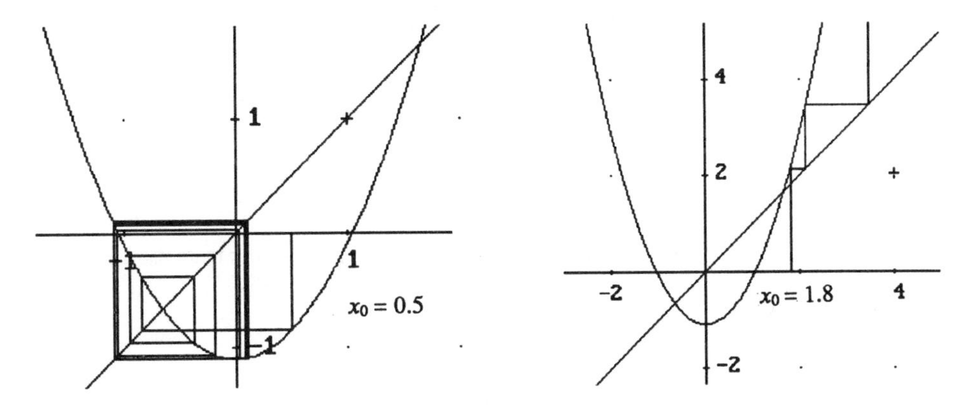

Fig 4.12 The system $x_{n+1} = x_n^2 - c$ for $c = 1.1$ with seed (a) $x_0 = 0.5$ and (b) $x_0 = 1.8$

TUTORIAL PROBLEM 4.12

Give an orbit analysis for the system

$$x_{n+1} = x_n^2 - 1.2$$

Demonstrate your answer by drawing cobweb diagrams.

The birth of a period-4 orbit

To obtain an insight into what happens as c moves through the critical value $c = \frac{5}{4}$ consider the graphs of $y = x$ and $y = F^2(x)$ for values of c just less than, equal to and greater than $\frac{3}{4}$. Figure 4.13 shows the transition from the point of intersection x_f^- which is stable $\left(c < \frac{3}{4}\right)$ to three points of intersection: x_f^- which is now unstable and x_p^- and x_p^+ which are stable. These form the period-2 cycle of $F(x)$.

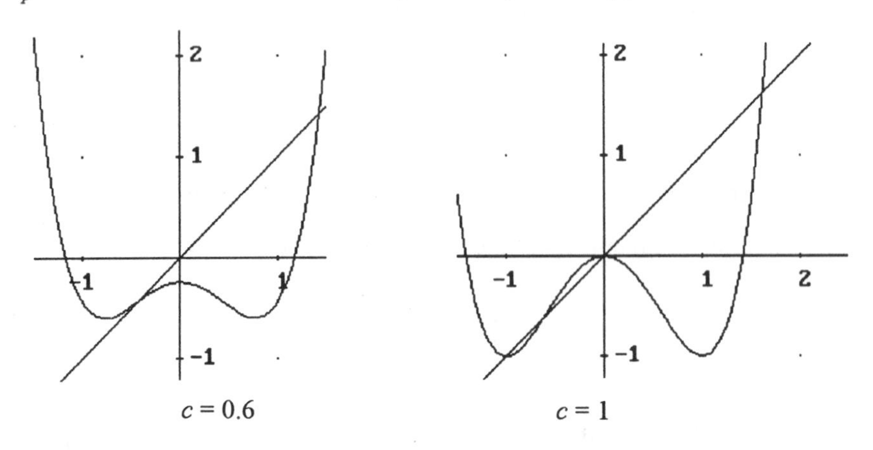

Fig 4.13 Graphs of $F^2(x)$ for $c < \frac{3}{4}$ and $c > \frac{3}{4}$

When $c = \frac{3}{4}$ the line $y = x$ is a tangent to the curve $y = F^2(x)$ and the transition to a period-2 orbit is about to occur.

As c increases the system changes as shown below:

	$c < \frac{3}{4}$		$c > \frac{3}{4}$
unstable	$x_f^+ = \frac{1}{2}(1 + \sqrt{1+4c})$	unstable	$x_f^+ = \frac{1}{2}(1 + \sqrt{1+4c})$
stable	$x_f^- = \frac{1}{2}(1 - \sqrt{1+4c})$	unstable	$x_f^- = \frac{1}{2}(1 - \sqrt{1+4c})$
		attracting periodic point	$x_p^+ = \frac{1}{2}(-1 + \sqrt{4c-3})$
		attracting periodic point	$x_p^- = \frac{1}{2}(-1 - \sqrt{4c-3})$

This process is known as **a bifurcation** and is illustrated in Fig 4.14. Along the horizontal axis are the values of c and along the vertical axis are the eventual values of the orbit for x_0 inside the fundamental interval.

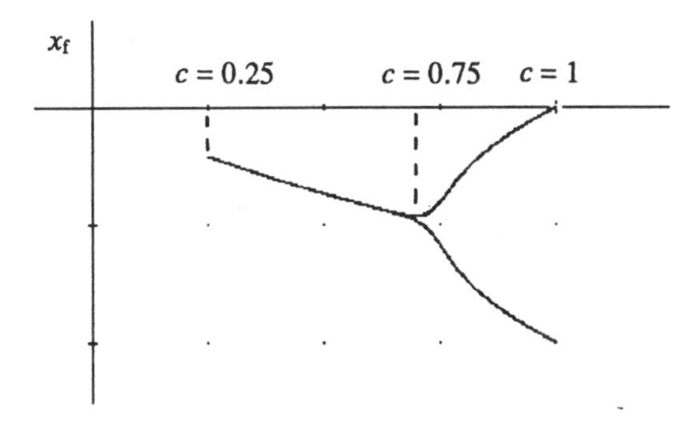

Fig 4.14 The bifurcation showing the birth of the period-2 cycle

When $c = \dfrac{5}{4}$ the period-2 cycle becomes repelling so we now have two unstable fixed points and a repelling period-2 cycle. Figure 4.15 shows graphs of $y = x$ and $y = F^4(x)$.

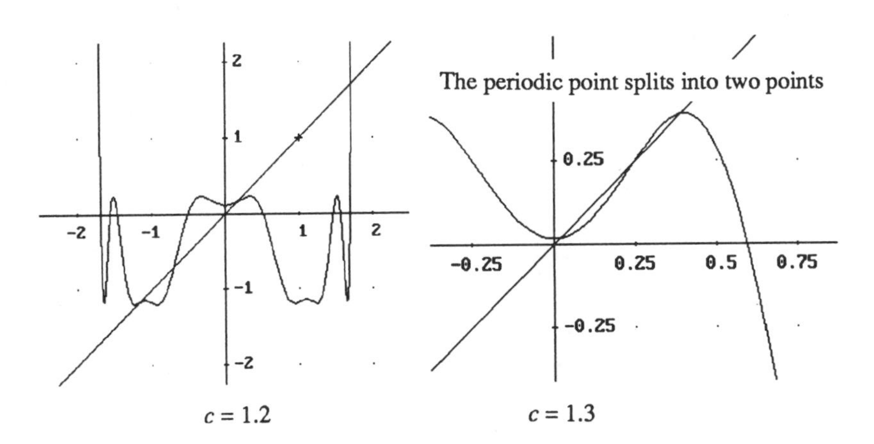

Fig 4.15 Graphs of $F^4(x)$ for $c = 1.2$ and $c = 1.3$

We see that the periodic points of period-2 have now each split into two points. We now have a period-4 cycle. The algebra for the analysis of the period-4 orbit becomes more demanding. $F^4(x)$ is a polynomial of degree 16 and so is the equation for the periodic points $F^4(x) = x$.

The following example shows the algebra that occurs in the analysis of period-4 orbits.

Example 8

Find the period-4 orbit for $x_{n+1} = x_n^2 - 1.3$.

Solution

We need to solve $F^4(x) = x$ for $F(x) = x^2 - 1.3$. Since the fixed points of $F^2(x)$ will also be solutions of $F^4(x) = x$, we begin by dividing $F^4(x) - x$ by $F^2(x) - x$ to give the polynomial equation of degree 12:

$$x^{12} - 7.8x^{10} + x^9 + 21.45x^8 - 5.2x^7 - 22.66x^6 + 7.54x^5 + 3.1655x^4$$

$$-1.028x^3 + 5.963\,62x^2 - 2.447\,89x + 0.045\,3180 = 0$$

A graph of this polynomial shows that there are four real solutions and an appropriate numerical method gives their values as −1.299 62, −1.148 66, 0.019 43, 0.389 018 correct to five decimal places. The period-4 orbit is

−1.299 62, 0.389 018, −1.148 66, 0.019 43

and is shown in Fig 4.16.

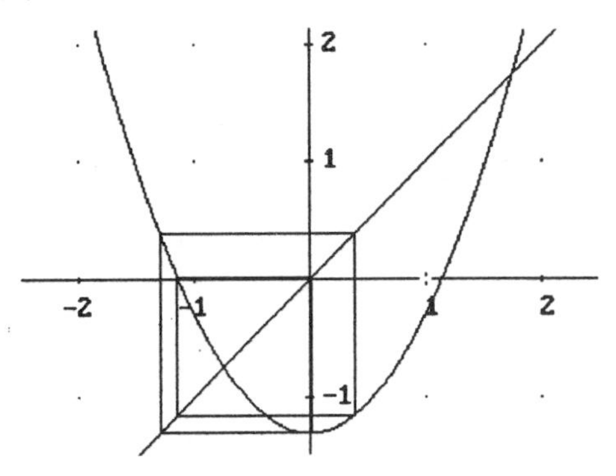

Fig 4.16 The period-4 orbit of $x_{n+1} = x_n^2 - 1.3$

TUTORIAL PROBLEM 4.13

Investigate the system $x_{n+1} = x_n^2 - 1.38$ and show that a period-8 cycle exists for this value of c.

To proceed further we need to adopt a numerical approach. Period doubling takes place for smaller increases in the value of c. It turns out that the sequence of values of critical values of c, c_k say, for 2^k cycles, converges to a limiting value, c_∞. Table 4.2 shows this sequence of c_k values and their difference.

k	Birth of:	c_k	$c_k - c_{k-1}$
1	2-cycle	0.75	–
2	4-cycle	1.25	0.5
3	8-cycle	1.368 099	0.118 099
4	16-cycle	1.394 046	0.025 947
5	32-cycle	1.399 631	0.005 585
6	64-cycle	1.400 830	0.001 199
7	128-cycle	1.401 085	0.000 255
8	256-cycle	1.401 140	0.000 055

Table 4.2 Sequence of critical c values

The limiting value of c_k is 1.401 155 to six decimal places. Much work on the quadratic map was carried out by the American mathematician Mitchell Feigenbaum in the 1970s. He found that the ratio $\dfrac{c_{k+1} - c_k}{c_k - c_{k-1}}$ tends towards a constant number $\delta = 4.669\ 202$ to six decimal figures. This is known as **Feigenbaum's constant**. It turns out that δ is a universal constant in that it appears whenever a non-linear system undergoes the period doubling bifurcations similar to the quadratic map.

Figure 4.17 shows the graph of bifurcations for $F(x) = x^2 - c$ as c increases from 0.75 towards the limiting value. To draw the figure, for each value of c we have calculated 400 iterations of the system and discarded the first 50 of them so that the iterative process has settled down. Then we plot the orbit for each value of c. This is often called the **orbit diagram** or **bifurcation diagram**. A lot more detail can be obtained by increasing the number of iterations.

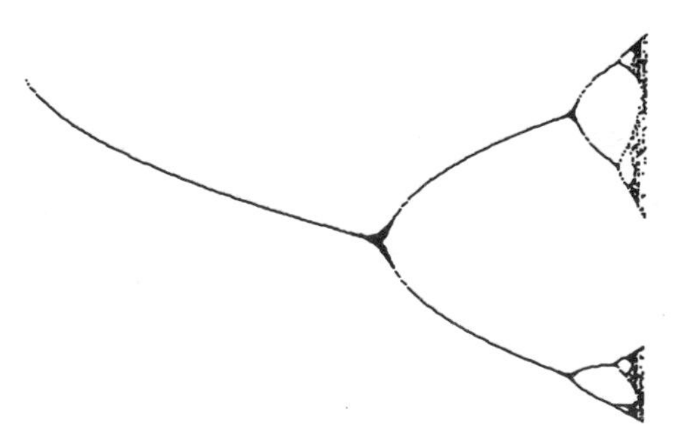

Fig 4.17 The orbit diagram for $0.25 < c < c_\infty$

As the period doubling occurs, a new 2^k cycle is formed which is attracting. After rejecting the first few hundred iterations the periodic orbit becomes apparent.

TUTORIAL PROBLEM 4.14

Investigate the system $x_{n+1} = x_n^2 - 1.4$ numerically. Find the periodic cycle. Show that the fundamental period is 32.

Beyond the limiting value c_∞

As we increase c beyond the limiting value c_∞, iterations have surprising behaviour with regions of chaotic behaviour and periodic cycles. Figure 4.18 shows cobweb diagrams for
$c = 1.5$ and $c = 1.75$ with $x_0 = 0$.

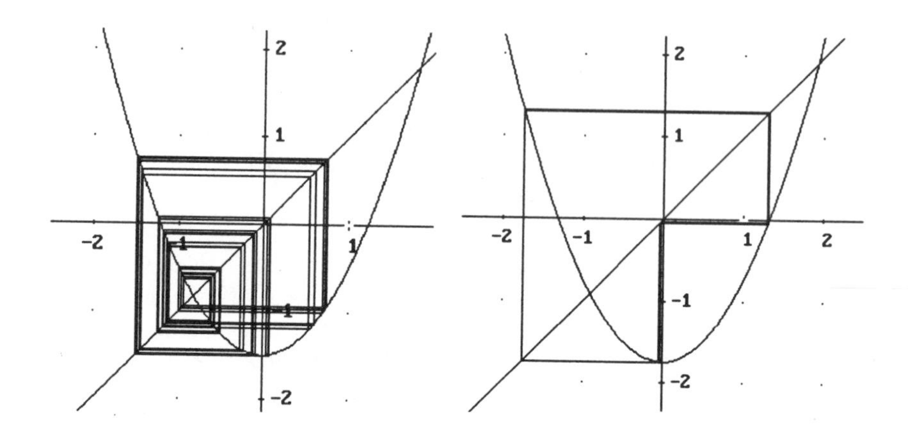

Fig 4.18 Cobweb diagrams for $F(x) = x^2 - 1.5$ (left) and $F(x) = x^2 - 1.75$ (right)

The system $x_{n+1} = x_n^2 - 1.5$ is an example of chaotic behaviour, with the iterations appearing to visit points within the fundamental interval in a random manner not attracted to any periodic cycle. However for $x_{n+1} = x_n^2 - 1.75$, order appears to have returned with the iterations attracted to a period-3 cycle.

A natural question to ask is if an initial point within the fundamental interval is captured by this interval for all values of c.

Example 9

Show that the condition for the fundamental interval $[-x_f^+, x_f^+]$ to be mapped onto itself by the map $F(x)$ is that $|F(0)| \le x_f^+$.

Find the range of values of c for the fundamental interval to be invariant.

Solution

The critical initial value is $x_0 = 0$. If $F(0)$ lies in the fundamental interval then the interval is mapped onto itself.

Suppose that $|F(0)| > x_f^+$; then the first iterate x_1 is outside the fundamental interval and the orbit diverges since x_f^+ is unstable. Figure 4.19 shows the cobweb diagram for such an orbit.

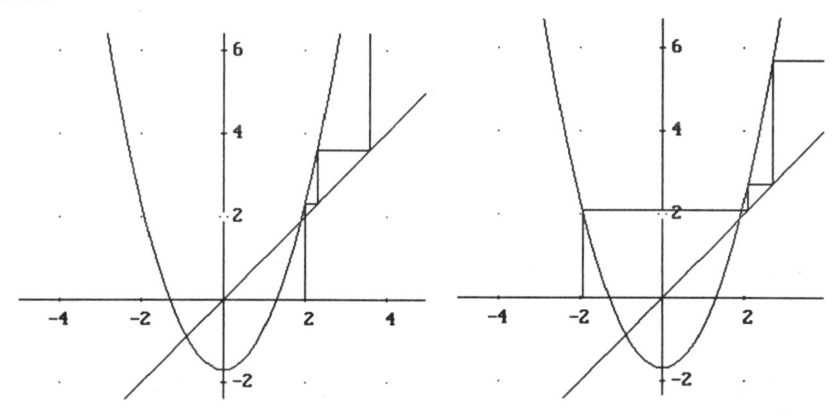

Fig 4.19 For $|F(0)| > x_f^+$ the orbit diverges

Suppose that $|F(0)| \le x_f^+$; then $|x_1| = |F(0)| \le x_f^+$. The next iterate $|x_2| = |F(x_1)| \le |F(x_f^+)|$ since $|x_1| \le |x_f^+|$ and so $|x_2| \le x_f^+$. Continuing these iterations shows that $|x_n| \le |x_f^+|$ for all n. The fundamental interval is invariant. Figure 4.20 shows the cobweb diagram for such an orbit.

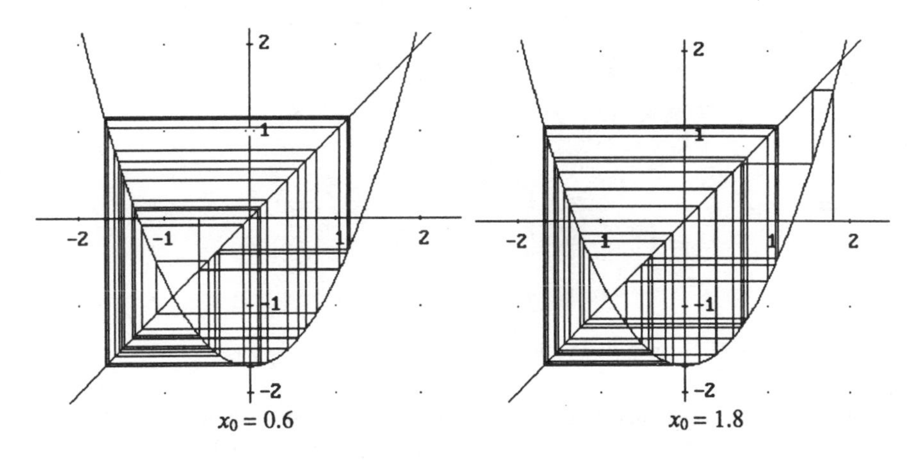

Fig 4.20 For $|F(0)| \le x_f^+$ the fundamental interval is invariant

The critical case occurs when $|F(0)| = x_f^+$. Now $x_f^+ = \frac{1}{2}(1+\sqrt{1+4c})$ and $F(0) = -c$. We need to solve the equation

$$c = \frac{1}{2}(1+\sqrt{1+4c})$$

for c.

$\Rightarrow \quad 2c - 1 = \sqrt{1+4c}$

Squaring both sides,

$\Rightarrow \quad (2c-1)^2 = 1 + 4c$

$\Rightarrow \quad 4c - 8c = 0$

$\Rightarrow \quad c = 0 \quad \text{or} \quad c = 2$

The solution $c = 0$ is a spurious solution introduced by the squaring process. The critical value of c is $c = 2$. We deduce that the fundamental interval is invariant for all values of c in the interval $c_\infty < c \leq 2$.

Figure 4.21 shows the extension of Fig 4.17 where the iterations are plotted after the first 50 iterations have been discarded.

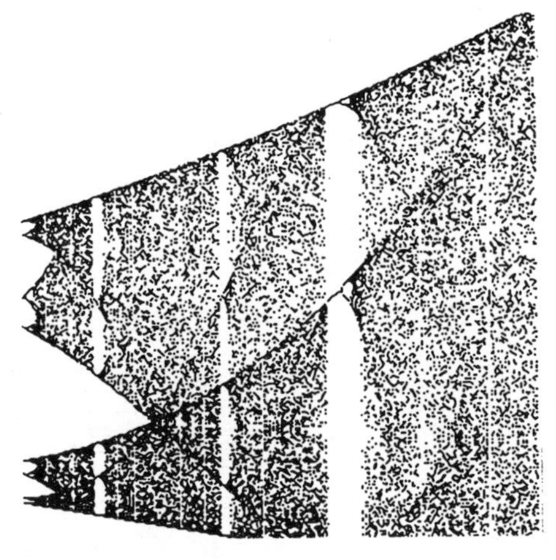

Fig 4.21 The orbit diagram for the quadratic map $F(x) = x^2 - c$ for $c_\infty < c \leq 2$

There are dark regions of chaotic behaviour followed by bands of white space where the iterations are attracted to periodic orbits. For example, around $c = 1.75$ there appears to be a period-3 cycle which then bifurcates to a period-6 cycle and so on.

The transition from period-2 to period-4 to period-8 cycles seems a smooth process and then beyond the critical value there seem to be many c-values where the system is chaotic. A remarkably simple theorem provides us with an important result.

The Period-3 Theorem

Suppose that $F(x)$ is a continuous map with a periodic point of prime period 3. Then F also has periodic points of all other periods.

What this theorem tells us is quite important. For some values of c the quadratic map seems to have one attracting cycle and no others, but if we can find a period-3 cycle then we are guaranteed that there are infinitely many other cycles for the quadratic map with every possible period. The Period-3 Theorem is a special case of Sarkovskii's Theorem first proved in 1964.

Sarkovskii's Theorem

Suppose that $F(x)$ is a continuous map. Define the natural numbers in the following order (known as the Sarkovskii ordering of the natural numbers):

$$3, 5, 7, 9, \ldots$$
$$2 \times 3, 2 \times 5, 2 \times 7, \ldots$$
$$2^2 \times 3, 2^2 \times 5, 2^2 \times 7, \ldots$$
$$2^3 \times 3, 2^3 \times 5, 2^3 \times 7, \ldots$$
$$\vdots$$
$$\ldots 2^n, 2^{n-1}, \ldots, 2^3, 2^2, 2, 1$$

Suppose that F has a periodic point of period n, where n precedes k in the Sarkovskii ordering. Then F has a periodic point of prime period k.

We have seen that if the quadratic map has a periodic point of period 2^5 then it has periodic points of period 2^4, 2^3, 2 and 1. The period-2^5 cycle is attracting and the others are repelling. We can now go further. If the quadratic map has a periodic point of period 28 ($= 2^2 \times 7$) then F also has periods 36, 44, 52, ..., 8, 4, 2, 1.

This section has only just started the story of the behaviour of chaotic systems. You have glimpsed the strange mixture of apparently random iterations to the structure of attracting periodic cycles. Most non-linear systems exhibit a similar orbit analysis to the quadratic map.

TUTORIAL PROBLEM 4.15

By solving $F^3(x) = x$, show that the quadratic map $F(x) = x^2 - 1.75$ has an attracting periodic 3-cycle. Find this period-3 cycle.

Confirm your result by drawing cobweb diagrams for various initial values x_0.

TUTORIAL PROBLEM 4.16

Use the program in Appendix A, or write your own, to investigate the orbit diagram for the map $F(x) = x^2 - c$ for values of c within (i) $1.25 < c < 1.41$, (ii) $1.6 < c < 1.65$, (iii) $1.75 < c < 1.8$.

Exercises 4.5

This exercise is a series of investigations that repeat the ideas of this section. You will need to explore the systems numerically as well as algebraically. Adapt your program from Tutorial Problem 4.16 to draw the orbit diagram.

1. Consider the logistic system

$$x_{n+1} = kx_n(1 - x_n), \quad k > 0$$

(i) Find the fixed points of the system. Classify them as stable or unstable.

(ii) For what values of k does a period-2 cycle exist?

(iii) Investigate the orbits as k increases from 0 to 4.

2. Show that a bifurcation of fixed points occurs for each of the following maps and $c = 1$. Draw the bifurcation diagram for values of c either side of $c = 1$.

(i) $F(x) = x^2 + x - c$

(ii) $F(x) = x^3 - cx$

(iii) $F(x) = c \sin c$

(iv) $F(x) = c(e^x - 1)$

3. Investigate the orbits of the non-linear system

$$x_{n+1} = c \sin x_n \qquad 0 \le x \le \pi$$

for $1 \le c \le \pi$.

4. Show that the quadratic map $F(x) = x^2 - 1.77$ has an attracting periodic 6-cycle. Find this period-6 cycle.

Confirm your result by drawing cobweb diagrams for various initial values x_0.

Further exercises

1. Consider the first order discrete system with iteration map

$$F(x) = \lambda x + (1 - \lambda)x^2$$

where λ is a parameter.

(i) Sketch the graph of $F(x)$. Find the fixed points of the system and show that for $\lambda \leq 4$ the fundamental interval is mapped onto itself.

(ii) Show that period doubling bifurcations occur for $\lambda \geq 3$. Find the period-2 cycles for $\lambda = 3.1$. Draw the cobweb diagram for some initial values for $\lambda = 3.1$.

(iii) Describe the process of period doubling for λ in the range $2 \leq \lambda < \lambda_\infty$, where λ_∞ is the limiting value for period doubling to occur.

2. For the quadratic map $F(x) = x^2 - c$ show that 0 is a point on a period-3 cycle if c satisfies the cubic equation

$$c^3 - 2c^2 + c - 1 = 0$$

Solve this equation for c and verify that it fits the bifurcation diagram in Fig 4.21.

3. Consider the dynamical system

$$x_{n+1} = 3.3x_n(1 - x_n)$$

(i) Find a period-2 cycle for this system.

(ii) Show that the 2-cycle is attracting by showing that

$$\left| \frac{dF}{dx}(p)\frac{dF}{dx}(q) \right| < 1$$

where the orbit is $p, q, p, q, \ldots$

4. Consider the dynamical system

$$x_{n+1} = 3.5x_n - 2.5x_n^2$$

(i) Show that $p = 0.535\ 947\ 556$ is one point on a period-4 cycle and hence find the period-4 orbit.

(ii) Show that this orbit is attracting. Verify your result by drawing cobweb diagrams for various initial values.

5. Consider the general 'tent map'

$$F(x) = \begin{cases} cx & 0 \le x \le \frac{1}{2} \\ c(1-x) & \frac{1}{2} < x \le 1 \end{cases}$$

where c is a parameter.

(i) Find and classify the fixed points of the system.

(ii) Sketch the orbit diagrams for $F(x)$ for values of c in the range $0 \le c \le 2$ and for $x_0 = \dfrac{1}{2}$. Describe each of your orbits.

(iii) Investigate where periodic cycles occur and the existence of bifurcations.

6. Consider the dynamical system

$$x_{n+1} = F(x_n) = x_n^3 - x_n^2 + (1-c)x_n + c$$

(i) Show that the fixed points are 1, $\sqrt{c}$ and $-\sqrt{c}$.

(ii) Find the values of c for which the fixed point $x_f = 1$ is attracting.

(iii) Show that the fixed point $x_f = -\sqrt{c}$ is repelling for $c > 0$.

(iv) Find the conditions on c for $x_f = \sqrt{c}$ to be attracting.

(v) Draw cobweb diagrams for various initial values for the two cases $c = -1$ and $c = 0.5$.

(vi) Show that there are bifurcations at $c = 0$ and $c = 1$. Draw the bifurcation diagram for $-2 < c < 1$.

The next two problems are extended activities which could form project work.

7. Compute the orbit diagrams corresponding to the initial value x_0 for the following maps:

$$A: \quad F(x) = cx(1-x), \quad 0 \le x \le 1,\ 1 \le c \le 4;\ x_0 = 0.5$$

$$B: \quad F(x) = a \sin x, \quad 0 \le x \le \pi,\ 0 \le a \le \pi;\ x_0 = \frac{\pi}{2}$$

Describe the similarities with the orbit diagram of the quadratic map $F(x) = x^2 - c$.

8. Consider the non-linear system

$$x_{n+1} = c \cos x_n$$

where c is a parameter.

(i) Compute the orbit diagram for this system for $0 \le c \le \pi$ using $x_0 = 0$ as the initial value.

(ii) By investigating regions of the orbit diagram to find intervals of c-values for which there is a single attracting fixed point, an attracting period-2 cycle, an attracting period-3 cycle and so on. Is there a period-3 cycle window?

5 •Projects

This chapter contains projects which draw on and extend the ideas developed in earlier chapters. The projects are presented as open-ended tasks so that students can explore them in different ways.

Project 1 The Geometric Chaos Game

Draw an equilateral triangle with vertices labelled A, B and C. Choose any point P_0 inside the triangle as the seed for the following iterative process:

- The point P_{n+1} is determined by randomly choosing a vertex and moving P_n halfway towards the chosen vertex. Mark the point with a dot.
- Repeat the process for many hundred (or thousand) points P_n building up the picture of the orbit of P_0.

Note that one way of randomly selecting the next vertex is to use a fair die. For example, let A represent 1 and 4, B represent 2 and 5, and C represent 3 and 6.

The image is an example of a geometric object called a **fractal**. Investigate what happens in the following cases:

1. The initial point P_0 is outside the triangle A, B, C.
2. The triangle ABC is not equilateral.
3. The initial shape is a square. What device could you use to randomly select the next vertex?

Project 2 Snowflakes and other Fractals

The following procedure describes the formulation of the snowflake curve.

Take an equilateral triangle, and on each side, at its midpoint, construct an equilateral triangle whose side is one-third the length of the side of the original triangle.

Now repeat this process, constructing an equilateral triangle on each straight-line segment of the resulting figure.

If the process were repeated indefinitely, investigate the length of the resulting outline and the area it would enclose.

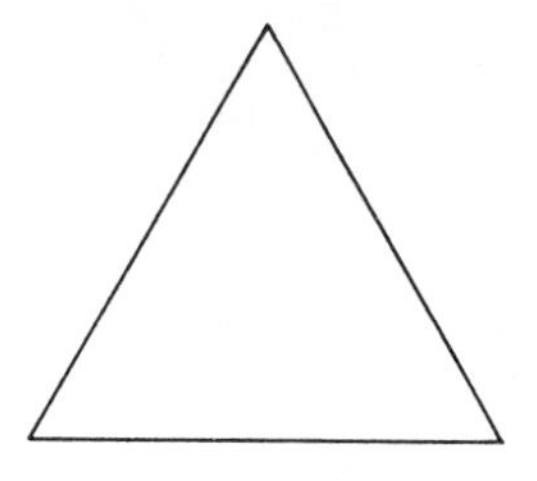

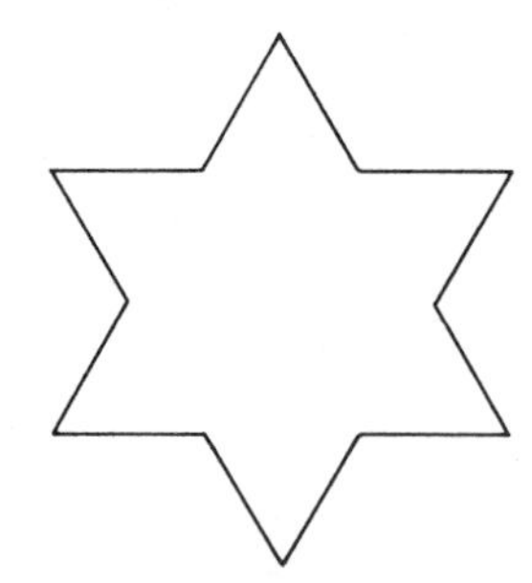

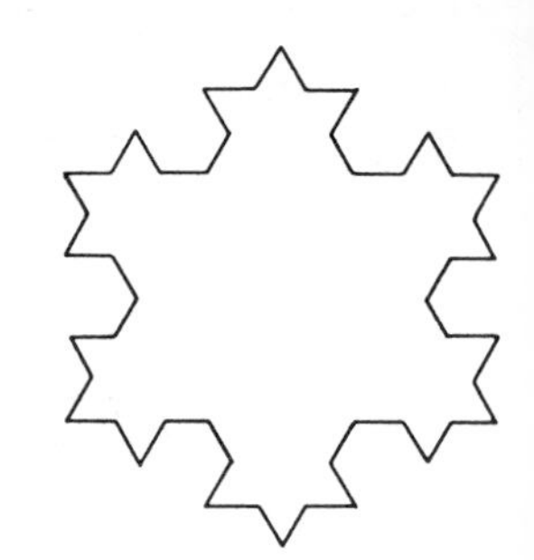

What happens if you start with other regular polygons?

Project 3 The Newton-Raphson Iterative Method

The Newton-Raphson method is an iterative scheme for solving an equation of the form $f(x) = 0$. The map for the method is

$$F(x) = x - \frac{f(x)}{f'(x)}$$

From what we have seen in Chapter 4, if $F(x)$ is non-linear we would expect chaotic behaviour.

1. What happens for $f(x) = 1 + x^2$?

2. Find the value of c so that the critical points for the map for

$$f(x) = (x^2 - 1)(x^2 + c)$$

lie on a period-2 cycle.

3. Investigate the behaviour of the fixed points when $f(x)$ is given by

$$f(x) = \frac{p(x)}{r(x)}$$

where $p(x)$ and $r(x)$ are polynomial functions. Classify the fixed points as attracting or repelling.

Illustrate your answers with particular choices of $p(x)$ and $r(x)$.

Project 4 Strange Pendulum Behaviour

The photograph shows an executive toy which can be obtained from novelty shops. It consists of a pendulum made up of two balls and two rings. In the base of the pendulum is a magnet. The motion of the system is unpredictable and chaotic.

A simple model of this toy consists of a small iron ball suspended over three magnets which form an equilateral triangle. If the ball is released it will be in motion because of magnetic forces and gravity, and will eventually settle over one of the magnets.

Formulate a model for the motion of the ball in the form of a pair of differential equations. By solving these equations numerically, investigate how the final resting place of the ball depends on its initial position.

Project 5 Lorenz Strange Attractors

The aim of this activity is to investigate the solutions of a simplified model for the atmosphere that led Lorenz to the unexpected behaviour that began the story of chaos.

The model differential equations of the Lorenz system are

$$\frac{dx}{dt} = -ax + ay$$

$$\frac{dy}{dt} = rx - y - xz$$

$$\frac{dz}{dt} = -bz + xy$$

where a, b and r are positive constants.

1. Show that for $0 < r \leq 1$ there is one fixed point at the origin 0; and that for $r > 1$ there are three fixed points 0, $(\alpha, \alpha, r-1)$, $(-\alpha, -\alpha, r-1)$, where $\alpha^2 = b(r-1)$.

2. Investigate the stability of these fixed points.

3. Choose values of $a = 10$ and $b = \frac{8}{3}$. Investigate numerical solutions of the Lorenz equations for various values of r in each of the following ranges: (i) $0 < r < 1$, (ii) $1 < r < 2$, and (iii) $2 < r < 30$.

Appendix A

DERIVE code for the activities in Chapter 4.

1. Generating the orbit of a sequence $x_{n+1} = F(x_n) = x_n^2 - 1$

   #1 Author F(x) := x^2 – 1
   #2 Author ITERATES(F(x),x,x_0,n) gives the sequence of n values starting with x_0

   Simplify

2. Forming $F^n(x)$ for $F(x) = x^2 - 1$

   #1 Author F(x) := x^2 – 1
   #2 Author F(F(x))
   Simplify gives $F^2(x)$
   #4 Author F(F(F(x)))
   Simplify gives $F^3(x)$

```
        2
#1:  F(x) := x  - 1

#2:  F(F(x))

      4     2
#3:  x  - 2·x

#4:  F(F(F(x)))

      8      6      4
#5:  x  - 4·x  + 4·x  - 1
```

```
COMMAND: Author Build Calculus Declare Expand Factor Help Jump soLve Manage
         Options Plot Quit Remove Simplify Transfer Unremove moVe Window approX
Compute time: 0.0 seconds                                           Derive XM
Simp(#4)                                 Free:100%                    Algebra
```

3. Cobweb and staircase diagrams

   #1 [[vsubr,vsub(r+1)],[vsub(r+1),vsub(r+1)],[vsub(r+1),vsub(r+2)]]
   #2 vector(#1,r,1,n–1)
   #3 [x,g,[[vsub1,0],[vsub1,vsub2]],#2]
   #4 COBWEB_AUX(v,g,x,n) := #3
   #5 COBWEB(g,x,a,n) := COBWEB_AUX(ITERATES(g,x,a,n),g,x,n)

   To run the program, first define a function using

   Author F(x) :=

   and then

Author COBWEB(F(x),x,a,n)
approXimate

It is important to choose actual values for a and n, as in the following screen dump.

$$\#4:\quad \text{COBWEB\_AUX}(v, g, x, n) := \left[x, g, \begin{bmatrix} v_1 & 0 \\ v_1 & v_2 \end{bmatrix}, \text{VECTOR}\left(\begin{bmatrix} v_r & v_{r+1} \\ v_{r+1} & v_{r+1} \\ v_{r+1} & v_{r+2} \end{bmatrix}, r, \right.\right.$$

```
#5:  COBWEB(g, x, a, n) := COBWEB_AUX(ITERATES(g, x, a, n), g, x, n)
```

$$\#6:\quad F(x) := x^2 - 1$$

```
#7:  COBWEB(F(x), x, 0.5, 10)
```

$$\#8:\quad \left[x, x^2 - 1, \begin{bmatrix} 0.5 & 0 \\ 0.5 & -0.75 \end{bmatrix}, \left[\begin{bmatrix} 0.5 & -0.75 \\ -0.75 & -0.75 \\ -0.75 & -0.4375 \end{bmatrix}, \begin{bmatrix} -0.75 & -0.4375 \\ -0.4375 & -0.4375 \\ -0.4375 & -0.808593 \end{bmatrix}, \left[\right.\right.\right.$$

```
COMMAND: Author Build Calculus Declare Expand Factor Help Jump soLve Manage
         Options Plot Quit Remove Simplify Transfer Unremove moVe Window approX
Compute time: 0.8 seconds                                          Derive XM
Approx(#7)                              Free:100%                    Algebra
```

4. Generating data points for the Bifurcation diagram.

```
#1   VECTOR([c,vSUBr],r,51,n)
#2   ORBIT(v,g,x,n,c) := #1
#3   ORB_DIAG(g,x,a,n,c) := ORBIT(ITERATES(g,x,a,n),g,x,n,c)
#4   f(c,n) := ORB_DIAG(x^2–c,x,0,n,c)
#5   LIST_F(list,n) := VECTOR(F(listSUBj,n),j,DIMENSION(list))
#6   L := VECTOR(CO+m*0.01,m,0,10)
#7   LIST_F(L,n)
```

In line #1 the value of n is the number of iterations of the map and this instruction ignores the first 50 iterations giving the map the opportunity to 'settle down'.

To run the program, first choose c and n and then Author f(c,n).
approX to give a sequence of n – 50 iterations.
Plot this sequence.
Repeating the process for many values of c produces the Bifurcation diagram.
Lines #5–#7 allow ten values of c to be iterated in one run of the program.

• Solutions to the Tutorial Problems

Chapter 1

Tutorial problem 1.1

The differential equation $\frac{dx}{dt} = a + bx$ is linear in the dependent variable x and is of the variables separable form. If we have a choice then the method of separation of variables is usually easier. Separating the variables and integrating,

$$\int \frac{1}{a+bx}dx = \int dt$$

$$\frac{1}{b}\ln(a+bx) = t + c \qquad b \neq 0$$

$$x = -\frac{a}{b} + Ae^{bt}$$

If $x = x_0$ when $t = 0$ then $A = x_0 + \frac{a}{b}$; so

$$x = -\frac{a}{b} + \left(x_0 + \frac{a}{b}\right)e^{bt}$$

For $b > 0$, $x \to \infty$ as $t \to \infty$.

For $b < 0$, $x \to -\frac{a}{b}$ as $t \to \infty$.

Chapter 2

Tutorial problems 2.1 and 2.2

The solutions for these problems are given in the text.

Tutorial problem 2.3

The graph and phase portrait of $v(x) = x(2 - x)$ is shown in the following figure.

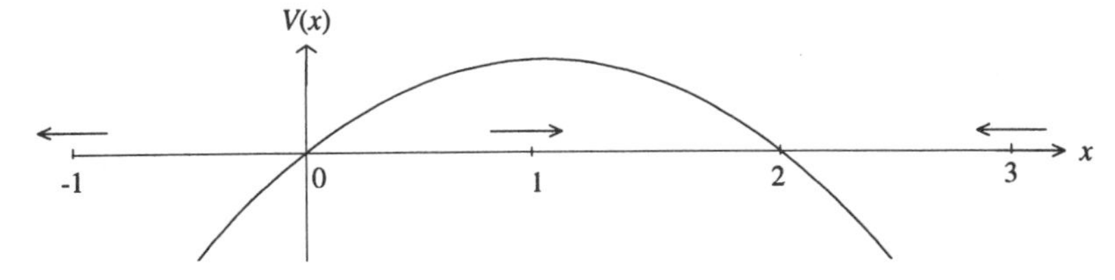

For $x_0 < 0$, $|v(x)|$ is increasing indefinitely.
For $0 < x_0 < 1$ the point gains speed until it reaches a maximum at $x = 1$ and then moves slower and slower as it approaches the fixed point at $x = 2$.
For $x_0 > 2$ the point moves slower and slower towards the fixed point at $x = 2$.
The invariant sets are $(-\infty,0)$, $(0,2)$, $(2,\infty)$.

Tutorial problem 2.4

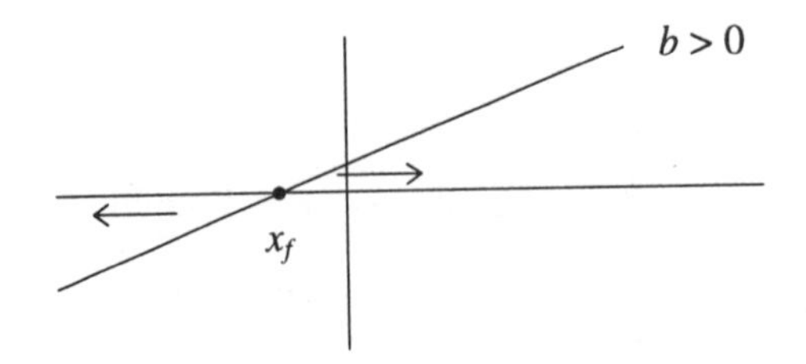

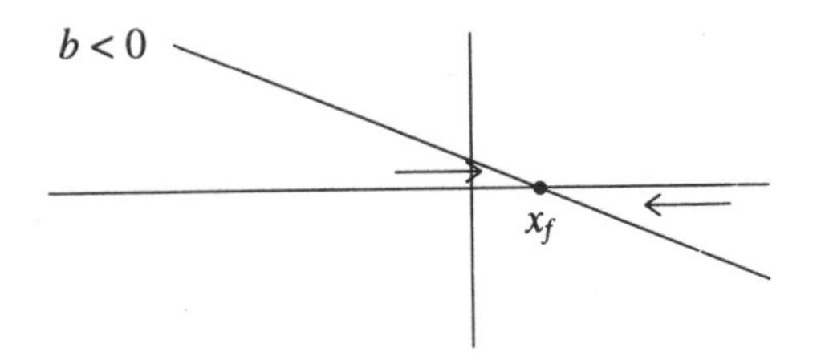

$b > 0$, $x_f = -\dfrac{a}{b}$ is a repellor.

$b < 0$, $x_f = -\dfrac{a}{b}$ is an attractor.

Tutorial problem 2.5

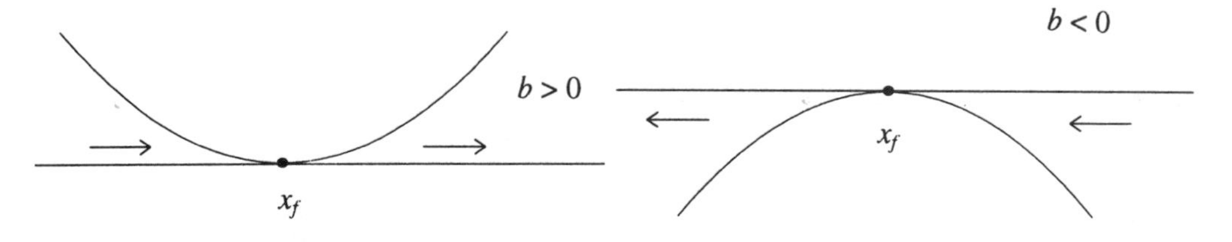

In each case the fixed point is not simple since $v'(x_f) = 0$.
The fixed points are examples of a shunt.
To solve the differential equation we separate the variables to give

$$\int \frac{1}{(x-x_f)^2}\,\mathrm{d}x = \int b\,\mathrm{d}t$$

$$-\frac{1}{(x-x_f)} = bt + c$$

With the initial condition $x = x_0$, $t = 0$ then $c = -\dfrac{1}{(x_0 - x_f)}$. Solving for x,

$$x - x_f = \frac{1}{\dfrac{1}{(x_0 - x_f)} - bt}$$

The solution is clearly not exponential.

Tutorial problem 2.6

Separating the variables,

$$\int \frac{1}{(x-x_f)^n}\,dx = \int dt$$

$$\frac{(x-x_f)^{-n+1}}{(-n+1)} = t + c, \qquad n \neq 1$$

$$t = 0,\ x = x_0 \quad \Rightarrow \quad c = \frac{(x_0 - x_f)^{-n+1}}{(-n+1)}$$

Hence $(x - x_f)^{-n+1} = (-n+1)t + (x_0 - x_f)^{-n+1}$.

For $n \geq 2$, $1 - n \leq -1 \Rightarrow x \rightarrow \infty$ in a finite time:

$$t = \frac{(x_0 - x_f)^{1-n}}{(n-1)} = \frac{1}{(n-1)(x_0 - x_f)^{n-1}}$$

Tutorial problem 2.7

Consider the motion near to the fixed point $x = -1$:

$$v(x) = \sqrt{1-x}\sqrt{1+x}$$

Now $\sqrt{1-x}$ is well behaved near $x = -1$ and it can be written as a Taylor polynomial expansion in the form

$$\sqrt{1-x} = \sqrt{2} - \frac{1}{2\sqrt{2}}(1+x) + 0((1+x)^2)$$

With this expansion $v(x)$ can be written as

$$v(x) = \sqrt{2}(1+x)^{\frac{1}{2}} - \frac{1}{2\sqrt{2}}(1+x)^{\frac{3}{2}} + 0(1+x)^{\frac{5}{2}}$$

If the initial condition is close to the natural boundary $x = -1$, then

$$\frac{dx}{dt} = v(x) \approx \sqrt{2}(1+x)^{\frac{1}{2}}$$

Solving for x we have

$$x = -1 + \left((1+x_0)^{\frac{1}{2}} + \frac{t}{\sqrt{2}}\right)^2$$

As t increases, x increases away from the natural boundary at $x = -1$.

To solve the equation of motion analytically,

$$\frac{dx}{dt} = \sqrt{1-x^2}$$

Let $x = \sin y$, so $\dfrac{dx}{dt} = \cos y \dfrac{dy}{dt}$. On substitution

$$\cos y \frac{dy}{dt} = \sqrt{1-\sin^2 y} = \cos y \quad \Rightarrow \quad \frac{dy}{dt} = 1$$

Integrating gives $y = t + c$.

Since $x = \sin y$, $x = \sin(t + c)$.

If $x = x_0$ when $t = 0$ then $c = \sin^{-1} x_0$, so

$$x = \sin(t + \sin^{-1} x_0)$$

On putting $x = 1$ we obtain the critical time at which the motion terminates:

$$t_c = \tfrac{1}{2}\pi - \sin^{-1} x_0$$

Tutorial problem 2.8

(i) The logistic model is

$$v(P) = aP\left(1 - \frac{P}{M}\right)$$

The fixed points are $P = 0$ and $P = M$. The phase portrait is

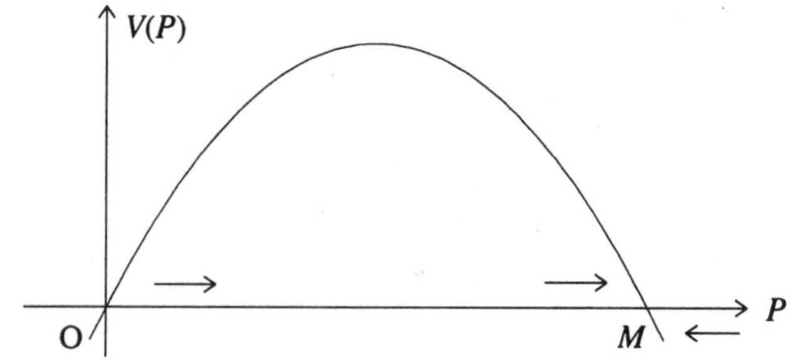

(a) For $P(0) = P_0 < M$ the population increases to the equilibrium value M.

(b) For $P(0) = P_0 > M$ the population decreases to the equilibrium value M.

(ii) For small P_0 the velocity field can be approximated by

$$v(P) = aP$$

Hence, $\dfrac{\mathrm{d}P}{\mathrm{d}t} = aP$ can be solved analytically to give

$$P = P_0 \mathrm{e}^{at}$$

(iii) For P_0 close to M we expand $v(P)$ as a Taylor polynomial about $P = M$ to give

$$v(P) \approx -a(P - M) + 0(P - M)^2$$

Hence, $\dfrac{\mathrm{d}P}{\mathrm{d}t} = -a(P - M)$ can be solved analytically to give

$$P = M + (P_0 - M)\mathrm{e}^{-at}$$

(iv) Start with

$$P = \frac{M}{1 + \left(\dfrac{M}{P_0} - 1\right)\mathrm{e}^{-at}}$$

For small P_0 we multiply the top and bottom by $\dfrac{P_0}{M}\mathrm{e}^{at}$ to give

$$P = \frac{P_0 \mathrm{e}^{at}}{1 + \dfrac{P_0}{M}(\mathrm{e}^{at} - 1)}$$

For small P_0 and t, $P \approx P_0\mathrm{e}^{at}$, thus confirming part (ii).

For P_0 close to M the factor $\varepsilon = \left(\dfrac{M}{P_0} - 1\right)\mathrm{e}^{-at}$ is small, so

$$P = \frac{M}{1+\varepsilon} = M(1 - \varepsilon + 0(\varepsilon^2))$$

$$\Rightarrow \quad P \approx M\left(1 - \left(\frac{M}{P_0} - 1\right)\mathrm{e}^{-at}\right)$$

$$\Rightarrow \qquad P \approx M - \frac{M}{P_0}(M - P_0)\mathrm{e}^{-at} \approx M + (P_0 - M)\mathrm{e}^{-at}$$

since $P_0 \approx M$, thus confirming part (iii).

Tutorial problem 2.9

(i) For $c < \dfrac{aM}{4}$ the velocity function has phase portrait.

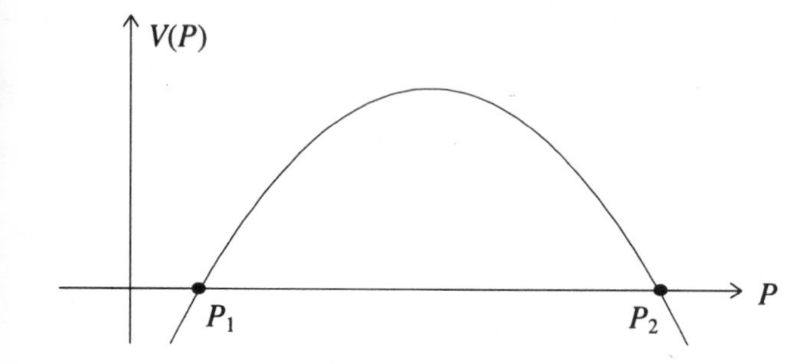

Note that $\dfrac{aM}{4}$ is the maximum value of $aP\left(1 - \dfrac{P}{M}\right)$.

The effect of the constant harvesting model is to translate the graph of $g(P) = aP\left(1 - \dfrac{P}{M}\right)$ downwards by an amount c. The new fixed points are

$$P_1 = \frac{M}{2}\left(1 - \sqrt{1 - \frac{4c}{a}}\right) \qquad \text{which is unstable}$$

$$P_2 = \frac{M}{2}\left(1 + \sqrt{1 - \frac{4c}{a}}\right) \qquad \text{which is stable}$$

(ii) (a) $0 < P_c < P_1$, the population is unstable and will become extinct.
(b) $P_1 < P_c < P_2$, the population will increase towards P_2.
(c) $P_2 < P_c$, the population will decrease towards P_2.

Chapter 3

Tutorial problem 3.1

(i) With $x_2 = \dfrac{\mathrm{d}x}{\mathrm{d}t}$ then $\dfrac{\mathrm{d}^2x}{\mathrm{d}t^2} = \dfrac{\mathrm{d}x_2}{\mathrm{d}t}$ and the differential equation becomes

$$\frac{\mathrm{d}x_2}{\mathrm{d}t} + x_2 + x_1 - x_1^3 = 0$$

The pair of first order equations is

$$\frac{dx_1}{dt} = x_2 \quad \text{and} \quad \frac{dx_2}{dt} = -x_1 + x_1^3 - x_2$$

(ii) Starting with $x_2 = \dfrac{dx}{dt} + x$ we have $\dfrac{dx_2}{dt} = \dfrac{d^2x}{dt^2} + \dfrac{dx}{dt}$ and the differential equation becomes

$$\frac{dx_2}{dt} + x_1 - x_1^3 = 0$$

The pair of first order equations is

$$\frac{dx_1}{dt} = x_2 - x_1 \quad \text{and} \quad \frac{dx_2}{dt} = x_1^3 - x_1$$

Tutorial problem 3.2

(i) The pair of first order equations can be transformed to the second order equation

$$\frac{d^2x_1}{dt^2} + 2\frac{dx_1}{dt} + 2x_1 = 2$$

Solving for x_1 gives the general solution

$$x_1 = e^{-t}(A \cos t + B \sin t) + 1$$

Substituting into the first equation for $\dfrac{dx_1}{dt}$ and x_1, and solving for x_2, gives

$$x_2 = e^{-t}(-A \sin t + B \cos t) + 2$$

The initial conditions $x_1(0) = 0$ and $x_2(0) = 3$ lead to $A = -1$ and $B = 1$. The particular solution is

$$x_1 = 1 + e^{-t}(-\cos t + \sin t)$$
$$x_2 = 2 + e^{-t}(\sin t + \cos t)$$

(ii) As $t \to \infty$, $x_1 \to 1$ and $x_2 \to 2$.

Tutorial problem 3.3

The figure overleaf shows the solution curves with directions added.

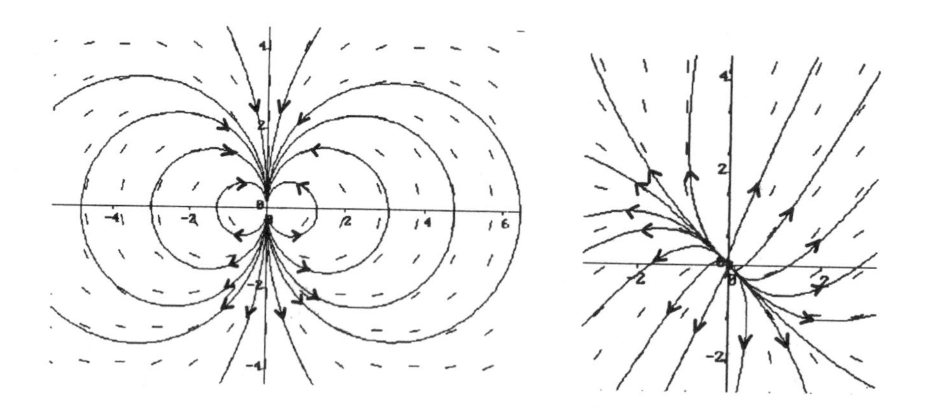

For system (i) an initial point $x_2 > 0$ is attracted towards the origin; for $x_2 < 0$ the system appears to be a repellor. For system (ii) any initial point moves away from the origin.

Tutorial problem 3.4

The fixed points are solutions of the equations

$$2 - x_1 x_2 = 0$$
$$x_2(1 - x_1) - 3 = 0$$

Substituting for $x_1 x_2$ from the first equation into the second

$$x_2 - 2 - 3 = 0 \quad \Rightarrow \quad x_2 = 5$$

From the first equation

$$x_1 x_2 = 2 \quad \Rightarrow \quad x_1 = \frac{2}{5}$$

The fixed point of the system is $\left(\frac{2}{5}, 5\right)$.

Tutorial problem 3.5

(i) Points A, C and D attract some motion.
(ii) Points A, B and C repel some motion.
(iii) A is unstable; B is strongly unstable; C is unstable; D is strongly stable.

Tutorial problem 3.6

(i) For $y = \frac{dx}{dt}$ we have $\frac{dy}{dt} = \frac{d^2x}{dt^2} = -x$.

The second order differential equation reduces to the pair

$$\frac{dx}{dt} = y \quad \text{and} \quad \frac{dy}{dt} = -x$$

(ii) Starting with $x(t) = x_0 \cos t + y_0 \sin t$ we have

$$\frac{dx}{dt} = -x_0 \sin t + y_0 \cos t = y$$

and $$\frac{dy}{dt} = -x_0 \cos t - y_0 \sin t = -x$$

Hence the given expressions for x and y satisfy the pair of first order differential equations.

With $t = 0$, $x(0) = x_0$ and $y(0) = y_0$, so the given expressions also satisfy the initial conditions.

(iii) Differentiate $x = r\cos\theta$ and $y = r\sin\theta$ with respect to t:

$$\dot{x} = \dot{r}\cos\theta - r\sin\theta\dot{\theta} = y = r\sin\theta \qquad (1)$$
$$\dot{y} = \dot{r}\sin\theta + r\cos\theta\dot{\theta} = -x = -r\cos\theta \qquad (2)$$

$(1) \times \cos\theta + (2) \times \sin\theta \Rightarrow \dot{r} = 0$
$(1) \times \sin\theta - (2) \times \cos\theta \Rightarrow \dot{\theta} = -1$

(iv) Solving for r and θ gives r = constant and $\theta = -t$ + constant. The solution curves are circles, centre the origin. Since θ decreases as t increases the motion is clockwise.

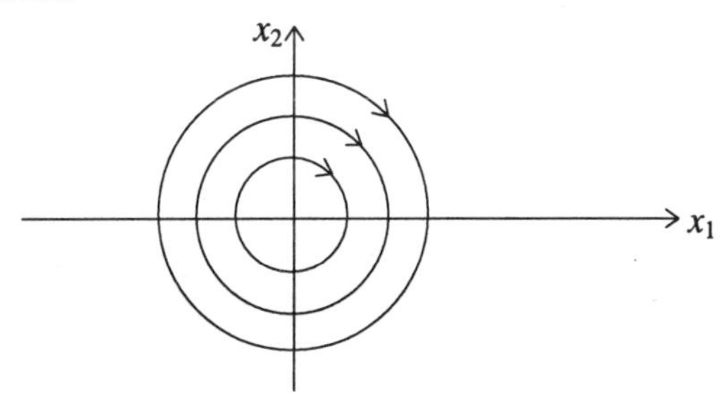

Tutorial problem 3.7

(i) The fixed point is given by

$$\mathbf{Ax} + \mathbf{h} = \mathbf{0}$$

Solving for $\mathbf{x}$, assuming that $\mathbf{A}^{-1}$ exists,

$$\mathbf{x}_f = -\mathbf{A}^{-1}\mathbf{h}$$

(ii) Let $\mathbf{X} = \mathbf{x} - \mathbf{x}_f$; then $\dfrac{\mathrm{d}\mathbf{x}}{\mathrm{d}t} = \dfrac{\mathrm{d}\mathbf{X}}{\mathrm{d}t}$. Substituting for $\mathbf{x}$ into the differential equation gives

$$\frac{\mathrm{d}\mathbf{X}}{\mathrm{d}t} = \mathbf{A}(\mathbf{X} + \mathbf{x}_f) + \mathbf{h}$$

$$= \mathbf{AX} + \mathbf{Ax}_f + \mathbf{h}$$

Now $\mathbf{Ax}_f + \mathbf{h} = \mathbf{0}$ is the definition of $\mathbf{x}_f$, so $\mathbf{X}$ satisfies the standard homogeneous form

$$\frac{\mathrm{d}\mathbf{X}}{\mathrm{d}t} = \mathbf{AX}$$

The geometrical significance of this transformation is a translation of the axes so that the fixed point moves to the origin in the new coordinate system.

Tutorial problem 3.8

(i) The eigenvalues are given by

$$\begin{vmatrix} 3-\lambda & 1 \\ 2 & 4-\lambda \end{vmatrix} = 0$$

$$(3-\lambda)(4-\lambda) - 2 = 0$$

$$\lambda^2 - 7\lambda + 10 = 0$$

$$(\lambda - 5)(\lambda - 2) = 0$$

The eigenvalues are $\lambda = 2$ and $\lambda = 5$. For $\lambda = 2$ the eigenvector is given by

$$\begin{bmatrix} 1 & 1 \\ 2 & 2 \end{bmatrix}\begin{bmatrix} u \\ v \end{bmatrix} = \mathbf{0}$$

$$u + v = 0$$

The eigenvector is $\alpha[1,-1]^{T}$ for any constant α.
For $\lambda = 5$, the eigenvector is given by

$$\begin{bmatrix} -2 & 1 \\ 2 & -1 \end{bmatrix}\begin{bmatrix} u \\ v \end{bmatrix} = \mathbf{0}$$

The eigenvector is $\beta[1,2]^{T}$ for any constant β.

Defining $\mathbf{P} = \begin{bmatrix} 1 & 1 \\ -1 & 2 \end{bmatrix}$ then $\mathbf{P}^{-1} = \frac{1}{3}\begin{bmatrix} 2 & -1 \\ 1 & 1 \end{bmatrix}$. Then

$$\begin{aligned} \mathbf{P}^{-1}\mathbf{A}\mathbf{P} &= \frac{1}{3}\begin{bmatrix} 2 & -1 \\ 1 & 1 \end{bmatrix}\begin{bmatrix} 3 & 1 \\ 2 & 4 \end{bmatrix}\begin{bmatrix} 1 & 1 \\ -1 & 2 \end{bmatrix} \\ &= \frac{1}{3}\begin{bmatrix} 2 & -1 \\ 1 & 1 \end{bmatrix}\begin{bmatrix} 2 & 5 \\ -2 & 10 \end{bmatrix} \\ &= \frac{1}{3}\begin{bmatrix} 6 & 0 \\ 0 & 15 \end{bmatrix} = \begin{bmatrix} 2 & 0 \\ 0 & 5 \end{bmatrix} \end{aligned}$$

(ii) The eigenvalues are given by

$$\begin{vmatrix} -\lambda & -1 \\ 4 & 4-\lambda \end{vmatrix} = 0$$

$$-\lambda(4-\lambda) + 4 = 0$$

$$\lambda^2 - 4\lambda + 4 = 0$$

$$(\lambda - 2)^2 = 0$$

The eigenvalue $\lambda = 2$ is repeated. The associated eigenvector is given by

$$\begin{bmatrix} -2 & -1 \\ 4 & 2 \end{bmatrix}\begin{bmatrix} u \\ v \end{bmatrix} = 0$$

$$-2u - v = 0$$

The eigenvector is $\alpha[1,-2]^{T}$ for any constant β.

Defining $\mathbf{P} = \begin{bmatrix} 1 & 1 \\ -2 & 0 \end{bmatrix}$ then $\mathbf{P}^{-1} = \frac{1}{2}\begin{bmatrix} 0 & -1 \\ 2 & 1 \end{bmatrix}$. Then

$$\mathbf{P}^{-1}\mathbf{AP} = \frac{1}{2}\begin{bmatrix} 0 & -1 \\ 2 & 1 \end{bmatrix}\begin{bmatrix} 0 & -1 \\ 4 & 4 \end{bmatrix}\begin{bmatrix} 1 & 1 \\ -2 & 0 \end{bmatrix}$$

$$= \frac{1}{2}\begin{bmatrix} 0 & -1 \\ 2 & 1 \end{bmatrix}\begin{bmatrix} 2 & 0 \\ -4 & 4 \end{bmatrix}$$

$$= \begin{bmatrix} 2 & 0 \\ 0 & 2 \end{bmatrix}$$

(iii) The eigenvalue equation is

$$\begin{vmatrix} 2-\lambda & 1 \\ -2 & 4-\lambda \end{vmatrix} = 0$$

$$(2-\lambda)(4-\lambda) + 2 = 0$$

$$\lambda^2 - 6\lambda + 10 = 0$$

$$\lambda = 3 \pm i$$

For complex eigenvalues and eigenvectors we will find an alternative transformation matrix **P** that is real.

Tutorial problem 3.9

(i) For a general matrix $\mathbf{A} = \begin{bmatrix} a & b \\ c & d \end{bmatrix}$ the eigenvalues are given by

$$\begin{vmatrix} a-\lambda & b \\ c & d-\lambda \end{vmatrix} = 0$$

$$(a-\lambda)(d-\lambda) - bc = 0$$

$$\lambda^2 - (a+d)\lambda + ad - bc = 0$$

Solving for λ we have

$$\lambda = \frac{1}{2}\left[(a+d) \pm \sqrt{(a+d)^2 - 4(ad-bc)}\right]$$

If $(a + d)^2 - 4(ad - bc) < 0$ then the eigenvalues are complex, $\lambda = \mu \pm i\omega$, where

$$\mu = \tfrac{1}{2}(a+d) \quad \text{and} \quad \omega = \tfrac{1}{2}\sqrt{4(ad-bc) - (a+d)^2}$$

(ii) Let $\mathbf{P} = \begin{bmatrix} a-\mu & -\omega \\ c & 0 \end{bmatrix}$; then $\mathbf{P}^{-1} = \dfrac{1}{c\omega}\begin{bmatrix} 0 & \omega \\ -c & a-\mu \end{bmatrix}$. Consider $\mathbf{P}^{-1}\mathbf{AP}$:

$$
\begin{aligned}
\mathbf{P}^{-1}\mathbf{AP} &= \frac{1}{c\omega}\begin{bmatrix} 0 & \omega \\ -c & a-\mu \end{bmatrix}\begin{bmatrix} a & b \\ c & d \end{bmatrix}\begin{bmatrix} a-\mu & -\omega \\ c & 0 \end{bmatrix} \\
&= \frac{1}{c\omega}\begin{bmatrix} 0 & \omega \\ -c & a-\mu \end{bmatrix}\begin{bmatrix} a(a-\mu)+bc & -a\omega \\ c(a-\mu)+cd & -c\omega \end{bmatrix} \\
&= \frac{1}{c\omega}\begin{bmatrix} 0 & \omega \\ -c & a-\mu \end{bmatrix}\begin{bmatrix} a(a-\mu)+bc & -a\omega \\ c\mu & -c\omega \end{bmatrix} \\
&= \frac{1}{c\omega}\begin{bmatrix} \mu c\omega & -c\omega^2 \\ c\omega^2 & \mu c\omega \end{bmatrix} \\
&= \begin{bmatrix} \mu & -\omega \\ \omega & \mu \end{bmatrix}
\end{aligned}
$$

as required.

Tutorial problem 3.10

The fixed points are solutions of the pair of equations

$$-x_1^2 + x_2 + 2 = 0 \qquad (1)$$

$$x_2^2 - x_1^2 = 0 \qquad (2)$$

From equation (2), $x_1 = x_2$ or $x_1 = -x_2$.
From equation (1),

$$x_1 = x_2 \quad \Rightarrow \quad -x_2^2 + x_2 + 2 = 0$$

$$(-x_2 + 2)(x_2 + 1) = 0$$

$$x_2 = 2 \text{ or } x_2 = -1$$

$$x_1 = -x_2 \quad \Rightarrow \quad x_2 = 2 \text{ or } x_2 = -1 \text{ also.}$$

There are four fixed points (–1,–1), (1,–1), (–2,2), (2,2).

Tutorial problem 3.11

The matrix of partial derivatives of the velocity is $\begin{bmatrix} -2x_1 & 1 \\ -2x_1 & 2x_2 \end{bmatrix}$.

At (–1,–1): $\mathbf{A} = \begin{bmatrix} 2 & 1 \\ 2 & -2 \end{bmatrix}$ tr($\mathbf{A}$) = 0 det($\mathbf{A}$) = –6

The fixed point is a saddle point.

At (1,–1): $\mathbf{A} = \begin{bmatrix} -2 & 1 \\ -2 & -2 \end{bmatrix}$ tr($\mathbf{A}$) = –4 det($\mathbf{A}$) = 6

The fixed point is a stable spiral.

At (–2,2): $\mathbf{A} = \begin{bmatrix} 4 & 1 \\ 4 & 4 \end{bmatrix}$ tr($\mathbf{A}$) = 8 det($\mathbf{A}$) = 12

The fixed point is an unstable node.

At (2,2): $\mathbf{A} = \begin{bmatrix} -4 & 1 \\ -4 & 4 \end{bmatrix}$ tr($\mathbf{A}$) = 0 det($\mathbf{A}$) = –12

The fixed point is a saddle point.

Tutorial problem 3.12

The fixed points are solutions of the pair of equations

$$x_1(2 - x_1) = 0$$
$$-x_2(1 - x_1) = 0$$

Solving for x_1 and x_2 gives two fixed points at (0,0) and (2,0).

The matrix of partial derivatives is $\mathbf{A} = \begin{bmatrix} 2-2x_1 & 0 \\ x_2 & x_1 - 1 \end{bmatrix}$.

At (0,0): $\mathbf{A} = \begin{bmatrix} 2 & 0 \\ 0 & -1 \end{bmatrix}$

The eigenvalues are 2 and –1; the fixed point is a saddle point.

At (2,0): $\mathbf{A} = \begin{bmatrix} -2 & 0 \\ 0 & 1 \end{bmatrix}$

The eigenvalues are –2 and 1; the fixed point is a saddle point.

The global phase portrait cannot be correct because it involves crossing phase curves which would lead to a non-unique solution at the point of intersection.

Tutorial problem 3.13

For polar coordinates let $x_1 = r\cos\theta$ and $x_2 = r\sin\theta$; then

$$\dot{x}_1 = \dot{r}\cos\theta - r\sin\theta\dot{\theta}$$

$$\dot{x}_2 = \dot{r}\sin\theta + r\cos\theta\dot{\theta}$$

Substituting for these into the two differential equations,

$$\dot{r}\cos\theta - r\sin\theta\dot{\theta} = r\sin\theta + r\cos\theta(1-r^2) \qquad (1)$$

$$\dot{r}\sin\theta + r\cos\theta\dot{\theta} = -r\cos\theta + r\sin\theta(1-r^2) \qquad (2)$$

$$(1)\times\cos\theta + (2)\times\sin\theta \quad \Rightarrow \quad \dot{r} = r(1-r^2)$$

$$(1)\times\sin\theta - (2)\times\cos\theta \quad \Rightarrow \quad -r\dot{\theta} = r \quad \Rightarrow \quad \dot{\theta} = -1$$

There is a limit cycle $r = 1$.
For $0 < r < 1$, $\dot{r} > 0$ so that r increases with t and an orbit converges towards the limit cycle.
For $r > 1$, $\dot{r} < 0$ so that r decreases with t and an orbit is attracted towards the limit cycle.

Tutorial problem 3.14

The model for a daily intake of 30μg is

$$\frac{dx_1}{dt} = -0.0361x_1 + 0.0124x_2 + 30$$

$$\frac{dx_2}{dt} = 0.0111x_1 - 0.0286x_2$$

The new equilibrium level is given by

$$-0.0361x_1 + 0.0124x_2 + 30 = 0$$

$$0.0111x_1 - 0.0286x_2 = 0$$

Solving for x_1 and x_2 we get

$$x_1 = 959\mu g, \quad x_2 = 372\mu g$$

(rounded to the nearest integer).

The matrix of coefficients of the system does not change and so the fixed point remains a stable node. The lead levels of 959μg in the blood and 372μg in the body tissues are stable.

For a daily intake of 70μg the new equilibrium level is

$$x_1 = 2237\mu g, \quad x_2 = 868\mu g$$

which are also stable.

Tutorial problem 3.15

Solving the three equations

$$\begin{aligned} -0.0361x_1 + 0.0124x_2 + 0.000\,035x_3 + 49.3 &= 0 \\ 0.0111x_1 - 0.0286x_2 \quad &= 0 \\ 0.0039x_1 - 0.000\,035x_3 &= 0 \end{aligned}$$

We obtain the equilibrium levels

$$x_1 = 1800\mu g, \quad x_2 = 699\mu g, \quad x_3 = 2 \times 10^5\,\mu g$$

The matrix of coefficients is

$$\mathbf{A} = \begin{bmatrix} -0.0361 & 0.0124 & 0.000\,035 \\ 0.0111 & -0.0286 & 0 \\ 0.0039 & 0 & -0.000\,035 \end{bmatrix}$$

The three eigenvalues for this matrix are –0.02, –0.045 and –0.000 031 (correct to 2 significant figures). Since they are all negative the system is stable.

Tutorial problem 3.16

The fixed points are soluitons of

$$\begin{aligned} (1 - aP_1 - P_2)P_1 &= 0 \\ (-1 - aP_2 + P_1)P_2 &= 0 \end{aligned}$$

Solving for P_1 and P_2 there are fixed points at (0,0), $\left(0, -\frac{1}{a}\right)$, $\left(\frac{1}{a}, 0\right)$ and $\left(\frac{1+a}{1+a^2}, \frac{1-a}{1+a^2}\right)$. The second fixed point $\left(-\frac{1}{a}, 0\right)$ is not physically realistic since $P_1 \geq 0$

and $P_2 \geq 0$.

The matrix of cofficients is

$$\mathbf{A} = \begin{bmatrix} 1-2aP_1 - P_2 & -P_1 \\ P_2 & -1-2aP_2 + P_1 \end{bmatrix}$$

At the non-trivial fixed point $\left(\dfrac{1+a}{1+a^2}, \dfrac{1-a}{1+a^2}\right)$.

$$\text{tr}(\mathbf{A}) = P_1 - P_2 - 2a(P_1 + P_2) = \frac{2a}{1+a^2} - 2a\left(\frac{2}{1+a^2}\right) = -\frac{2a}{(1+a^2)}$$

$$\det(\mathbf{A}) = (1-2aP_1 - P_2)(-1-2aP_2 + P_1) + P_1 P_2 = \frac{1-a^2}{1+a^2}$$

If $a = 0$ the linearized system has a centre at (1,1).
If $0 < a < 1$ then tr(**A**) < 0, det(**A**) > 0 and the fixed point is stable. Note that if $a > 1$ then det(**A**) < 0 and the fixed point becomes an (unstable) saddle.

For a spiral point, $\text{tr}(\mathbf{A})^2 - 4\det(\mathbf{A}) < 0$. Thus

$$\text{tr}(\mathbf{A})^2 - 4\det(\mathbf{A}) = \frac{4a^2 - 4(1-a^2)(1+a^2)}{(1+a^2)^2} < 0$$

$$\Rightarrow \quad a^4 + a^2 - 1 < 0$$

$$\Rightarrow \quad a^2 < \frac{\sqrt{5}-1}{2}$$

The fixed point is a stable spiral for $0 < a < \sqrt{\dfrac{\sqrt{5}-1}{2}} \approx 0.786$.

Chapter 4

Tutorial problem 4.1

(i) $x_{n+1} = 1.0325x_n$
$x_0 = 20\ 000$
At 25, $n = 13$, $x_n = £30\ 311$

(ii) $x_{n+1} = 1.005\,42x_n$
$x_0 = 20\,000$
At 25, $n = 78$, $x_n = £30\,464.80$

(iii) $x_{n+1} = 1.001\,25x_n$
$x_0 = 20\,000$
At 25, $n = 338$
$x_{n+1} = £30\,507.30$

If the interest is compounded m times per year then the total number of interest payments after t years will be mt.

The value of the investment will be

$$x_t = 20\,000\left(1+\frac{0.065}{m}\right)^{mt}$$

For continuous compounding $m \to \infty$,

$$\begin{aligned} x_t &= 20\,000 \lim_{m\to\infty}\left(1+\frac{0.065}{m}\right)^{mt} \\ &= 20\,000 \lim_{m\to\infty}\left(1+\frac{0.065t}{mt}\right)^{mt} \\ &= 20\,000\mathrm{e}^{0.065t} \end{aligned}$$

using the definition of the exponential function.

Applying this rule for 6.5 years gives an amount £30 515.40.

Tutorial problem 4.2

Using the same ideas as in Chapter 2, page 39, the increase in population during one unit of time is $P_{n+1} - P_n$. The logistic model gives

$$P_{n+1} - P_n = P_n g(P_n)$$

where g is a linear decreasing function of P_n. In general let $g(P_n) = a - bP_n$. Hence

$$P_{n+1} - P_n = P_n(a - bP_n)$$

$$P_{n+1} = P_n(1+a-bP_n) = (1+a)P_n\left(1-\frac{b}{(1+a)}P_n\right)$$

Let $M = \dfrac{1+a}{b}$ be the equilibrium population and $(1 + a) = k$; then

$$P_{n+1} = kP_n\left(1 - \frac{P_n}{M}\right)$$

If we write $\dfrac{P_n}{M} = F_n$ then F_n is the fraction of the equilibrium population level, so that $0 \le F_n = \dfrac{P_n}{M} \le 1$. F_n satisfies the difference equation

$$F_{n+1} = kF_n(1 - F_n)$$

$k = 1 + a$, where a is the growth rate for small populations.

Tutorial problem 4.3

(i) $F(x) = \sqrt{x}$

(ii) $256, 16, 4, 2, 2^{\frac{1}{2}}, 2^{\frac{1}{4}}, 2^{\frac{1}{8}}, 2^{\frac{1}{16}}$
256, 16, 4, 2, 1.4142, 1.1892, 1.0905, 1.0443

(iii) $F^2(x) = \sqrt{F(x)} = x^{\frac{1}{4}};\ (F(x))^2 = x$
$F^3(x) = \sqrt{F^2(x)} = x^{\frac{1}{8}};\ (F(x))^3 = x^{\frac{3}{2}}$
$F^4(x) = \sqrt{F^3(x)} = x^{16};\ (F(x))^4 = x^2$

Tutorial problem 4.4

$F(x) = -3x^2 + 2.5x + 0.5$
$F(1) = 0$
$F(0) = 0.5$
$F(0.5) = 1$
The period-3 cycle is 0, 0.5, 1.

$$F^2(x) = -3(F(x))^2 + 2.5F(x) + 0.5$$
$$= -27x^4 + 45x^3 - 17.25x^2 - 1.25x + 1$$
$$F^3(x) = F(F^2(x))$$
$$= -3(F^2(x))^2 + 2.5(F^2(x)) + 0.5$$
$$= -2187x^8 + 7290x^7 - 8869.5x^6 + 4455x^5 - 460.687x^4 - 286.875x^3 + 55.6875x^2 + 4.375x$$

$F^3(x) = x$ gives the 3-cycle.

Tutorial problem 4.5

(i) Let $x_n = \lambda^n$ so that $x_{n+1} = \lambda^{n+1}$. Thus

$$\lambda^{n+1} = b\lambda^n \; .$$

Solving for λ, $\lambda = b$. The general solution for the difference equation is $x_n = Ab^n$. Since $x = x_0$ when $n = 0$, $A = x_0$ and therefore

$$x_n = x_0 b^n$$

(ii) If $b = 1$ then $x_n = x_0$ for all n so that x_0 is a fixed point for all choices of x_0. If $b \neq 1$ then $x \neq bx$ for any x except $x = 0$. This is then the only fixed point.

(iii) If $|b| > 1$ then x_n grows indefinitely so that $x_f = 0$ is an unstable fixed point. If $|b| < 1$ then $x_n \to 0$ as $n \to \infty$ and $x_f = 0$ is stable.

Tutorial problem 4.6

The fixed points are solutions of the equation

$$x = \frac{1}{a+x}$$

$$\Rightarrow \qquad x_f = \frac{-a \pm \sqrt{a^2+4}}{2}$$

For $x_0 \geq 0$ there is one fixed point

$$\Rightarrow \qquad x_f^+ = \frac{-a + \sqrt{a^2+4}}{2}$$

With $F(x) = \dfrac{1}{a+x}$, $F'(x) = -\dfrac{1}{(a+x)^2}$.

For $x_f^+ = \dfrac{-a+\sqrt{a^2+4}}{2}$, $F'(x_f^+) = -x_f^{+2}$. Now $x_f^+ < 1$, so $\left|F'(x_f^+)\right| < 1$; thus this fixed point is stable.

Tutorial problem 4.7

(i) Starting with $x_0 = 0$, the orbit is given by 0, 1, 2, 0, 1, 2, ... so 0 is on the period-3 cycle 0, 1, 2.

(ii) For $x_0 = \frac{1}{3}$ the orbit is $\frac{1}{3}, \frac{5}{3}$, 1, 2, 0, ...
For $x_0 = \frac{4}{3}$ the orbit is $\frac{4}{3}, \frac{5}{3}$, 1, 2, 0, ...
For $x_0 = \frac{5}{3}$ the orbit is $\frac{2}{3}$, 2, 0, 1, ...

(iii) If α is a periodic point for the system $x_{n+1} = F(x_n)$ then solving $F(x) = \alpha$ will give a point that is eventually periodic. Similarly solving $F(x) = \beta$ if β is a fixed point will give an eventually fixed point.

Tutorial problem 4.8

(i) To differentiate $F^2(x)$ we use the chain rule:

$$\frac{\mathrm{d}}{\mathrm{d}x}F^2(x) = \frac{\mathrm{d}}{\mathrm{d}x}(F(F(x)) = \frac{\mathrm{d}}{\mathrm{d}F}(F(F(x)))\frac{\mathrm{d}F}{\mathrm{d}x}$$

In particular for the period-2 cycle x_0, x_1

$$\frac{\mathrm{d}F^2}{\mathrm{d}x}(x_0) = \frac{\mathrm{d}}{\mathrm{d}F}(F(F(x_0)))\frac{\mathrm{d}F}{\mathrm{d}x}(x_0) = \frac{\mathrm{d}F}{\mathrm{d}x}(x_1)\frac{\mathrm{d}F}{\mathrm{d}x}(x_0)$$

since $x_1 = F(x_0)$.

(ii) For $F(x) = x^2 - 1$, $\frac{\mathrm{d}F}{\mathrm{d}x} = 2x$, $x_0 = 0$ and $x_1 = -1$. So

$$\frac{\mathrm{d}F^2}{\mathrm{d}x}(x_0) = \frac{\mathrm{d}F}{\mathrm{d}x}(x_1)\frac{\mathrm{d}F}{\mathrm{d}x}(x_0) = (2\times -1)(2\times 0) = 0$$

Since $\frac{\mathrm{d}F^2}{\mathrm{d}x}(0) = 0$ the cycle is attracting.

(iii) Similarly applying the chain rule,

$$\frac{\mathrm{d}F^3}{\mathrm{d}x}(x_0) = \frac{\mathrm{d}}{\mathrm{d}x}F(F^2(x_0))\frac{\mathrm{d}}{\mathrm{d}F}F(F^2(x_0))\frac{\mathrm{d}F^2}{\mathrm{d}x}(x_0) = \frac{\mathrm{d}F}{\mathrm{d}x}(x_2)\frac{\mathrm{d}F}{\mathrm{d}x}(x_1)\frac{\mathrm{d}F}{\mathrm{d}x}(x_0)$$

(iv) For $F(x) = -\frac{3}{2}x^2 + \frac{5}{2}x + 1$, $\frac{\mathrm{d}F}{\mathrm{d}x} = -3x + \frac{5}{2}$, $x_0 = 0$, $x_1 = 1$ and $x_2 = 2$.

$$\frac{\mathrm{d}F}{\mathrm{d}x}(x_2) = -\frac{7}{2},\quad \frac{\mathrm{d}F}{\mathrm{d}x}(x_1) = -\frac{1}{2},\quad \frac{\mathrm{d}F}{\mathrm{d}x}(x_0) = \frac{5}{2}$$

So $$\frac{\mathrm{d}^3F}{\mathrm{d}x} = \left(-\frac{7}{2}\right)\left(-\frac{1}{2}\right)\left(\frac{5}{2}\right) = \frac{35}{8} > 1$$

The period-3 cycle is repelling. The cobweb diagram is shown below.

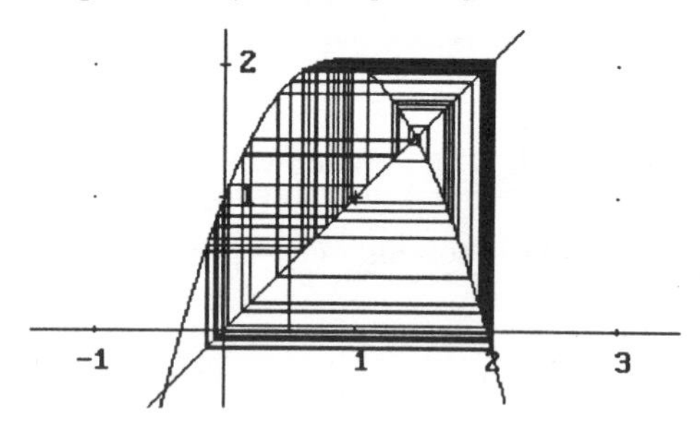

Tutorial problem 4.9

Let $x_n = \lambda + \mu y_n$ so that $x_{n+1} = \lambda + \mu y_{n+1}$.

Substitute for x_n and x_{n+1} into equation (4.2):

$$\lambda + \mu y_{n+1} = \alpha + \beta(\lambda + \mu y_n) + \gamma(\lambda + \mu y_n)^2$$

$$= (\alpha + \beta\lambda + \gamma\lambda^2) + (\beta\mu + 2\gamma\lambda\mu)y_n + \gamma\mu^2 y_n^2$$

$$y_{n+1} = \frac{\alpha + \lambda(\beta - 1) + \gamma\lambda^2}{\mu} + (\beta + 2\gamma\lambda)y_n + \gamma\mu y_n^2$$

Choose $\gamma\mu = 1$ and $\beta + 2\gamma\lambda = 0$. This gives

$$y_{n+1} = \frac{\alpha + \lambda(\beta - 1) + \gamma\lambda^2}{\mu} + y_n^2$$

Put $\mu = \dfrac{1}{\gamma}$ and $\gamma\lambda = -\dfrac{\beta}{2}$; then

$$y_{n+1} = \alpha\gamma - \tfrac{1}{4}\beta^2 + \tfrac{1}{2}\beta + y_n^2$$

Choose $c = -\alpha\gamma + \frac{1}{4}\beta^2 - \frac{1}{2}\beta$ to give the standard quadratic form

$$y_{n+1} = y_n^2 - c$$

Tutorial problem 4.10

(i) With $c = 0$ the map is $F(x) = x^2$. There are two fixed points.
$x_f = 0$ is a stable fixed point.
$x_f = 1$ is an unstable fixed point.
Note that $x_0 = -1$ is an eventually fixed point.

(ii) (a) With $c = -1$ the map is $F(x) = x^2 + 1$. The iteration with seed $x_0 = 0$ is 0, 1, 2, 5, 26, ... The sequence diverges for all x_0.

(b) With $c = 1$ the map is $F(x) = x^2 - 1$. The iteration with seed $x_0 = 0$ is 0, −1, 0, −1, ... This is a period-2 cycle. There are two fixed points found by solving $x = x^2 - 1$. We have $x_f = \frac{1}{2}(1+\sqrt{5})$ and $x_f = \frac{1}{2}(1-\sqrt{5})$.

Since $\left|\frac{dF}{dx}(x_f)\right| > 1$ for each x_f both fixed points are repelling.

Tutorial problem 4.11

For $F(x) = x^2 - \frac{3}{4}$ there are fixed points at $x_f = \frac{3}{2}$ and $x_f = -\frac{1}{2}$.

At $x_f = \frac{3}{2}$, $\dfrac{dF}{dx} = 3 > 1$ and the fixed point is unstable.

At $x_f = -\frac{1}{2}$, $\dfrac{dF}{dx} = -1$ and the fixed point is stable.

A period-2 cycle is about to be born as the fixed point near $x_f = -\frac{1}{2}$ changes from stable to unstable with c changing from just less than $\frac{3}{4}$ to just greater than $\frac{3}{4}$.

For any seed in the fundamental interval $\left[-\frac{3}{2}, \frac{3}{2}\right]$ the sequence converges to $x_f = -\frac{1}{2}$; for any seed outside the fundamental interval the sequence diverges.

Tutorial problem 4.12

Unstable fixed points at $\frac{1}{2}(1+5\sqrt{145})$ and $\frac{1}{2}(1-5\sqrt{145})$, i.e. 1.704 16, −0.704 16.

Period-2 cycle: $-\frac{1}{2}(1-0.6\sqrt{5})$, $-\frac{1}{2}(1+0.6\sqrt{5})$ which is attracting.

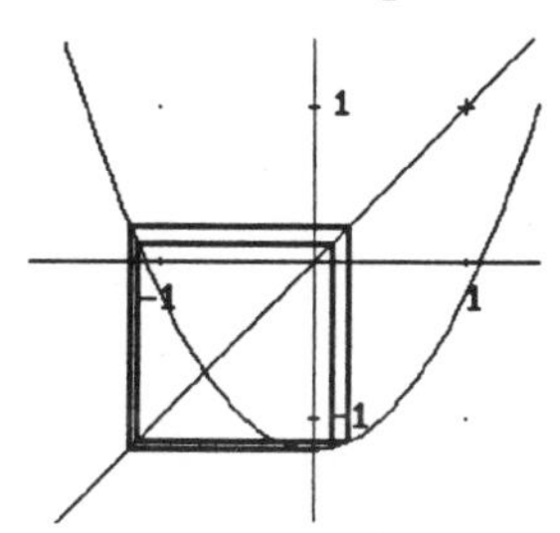

Tutorial problem 4.13

The screen dumps show the cobweb diagram with seed $x_0 = 0$.

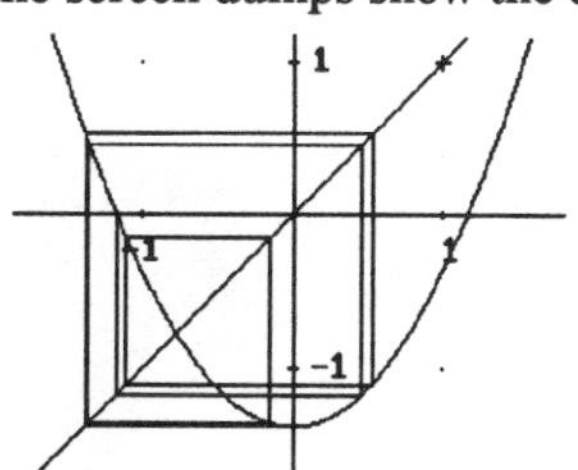

The period-8 cycle is −1.172 579, −0.005 058, −1.379 974, 0.524 329, −1.105 079, −0.158 801, −1.354 782, 0.455 435. This has been obtained by iterating 200 times with seed $x_0 = 0$; the sequence has then settled into a pattern. The function $F^8(x) - x$ is a polynomial of order 256 which is hard to factorize!

Tutorial problem 4.14

This problem requires some investigation so we do not give a specific solution. You may find the period-32 cycle to 4dp as −1.3999, 0.5599, −1.0864, −0.2197, −1.3517, 0.4272, −1.2175, 0.0823, −1.3932, 0.5411, −1.1072, −0.1741, −1.3697, 0.4761, −1.1733, −0.0233, −1.3995, 0.5585, −1.0881, −0.2161, −1.3533, 0.4315, −1.2138, 0.0734, −1.3946, 0.5450, −1.1030, −0.1833, −1.3664, 0.4670, −1.1819, −0.0031.

Tutorial problem 4.15

Since the fixed points are also period-3 cycles we solve $F^3(x) - x = 0$ by first dividing by $F(x) - x$ to give

$$\frac{F^3(x)-x}{F(x)-x} = x^6 + x^5 - 4.25x^4 - 2.5x^3 + 4.9375x^2 + 0.5625x + 0.015\,625$$

Setting this polynomial to zero and solving for x gives the period 3-cycle: −0.0549 5813, −1.746 979 60, 1.301 937 74.
The period-3 cycle is attracting and about to become a period-6 cycle because

$$\frac{dF}{dx}(x_0).\frac{dF}{dx}(x_1).\frac{dF}{dx}(x_2) = 8x_0x_1x_2 = 1$$

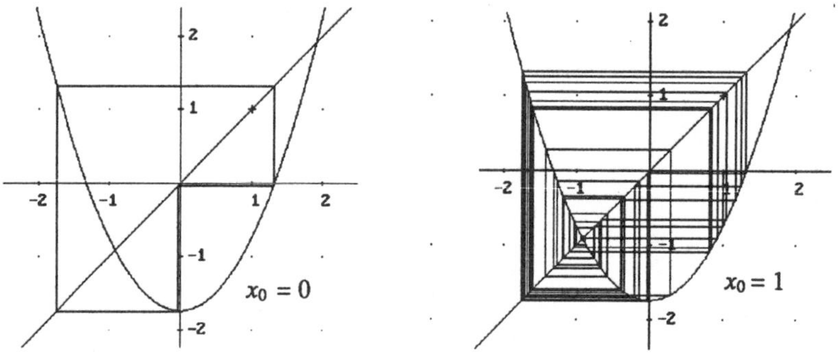

$x_0 = 0$

$x_0 = 1$

Tutorial problem 4.16

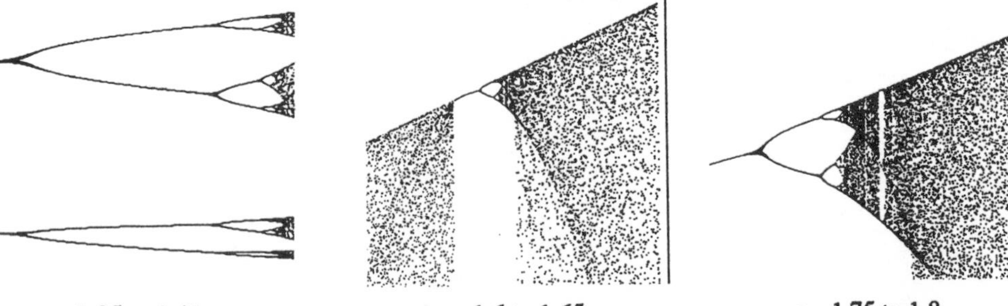

$c = 1.25$ to 1.41 $c = 1.6$ to 1.65 $c = 1.75$ to 1.8

• Answers to the Exercises

Chapter 1

Exercises 1.1

1.

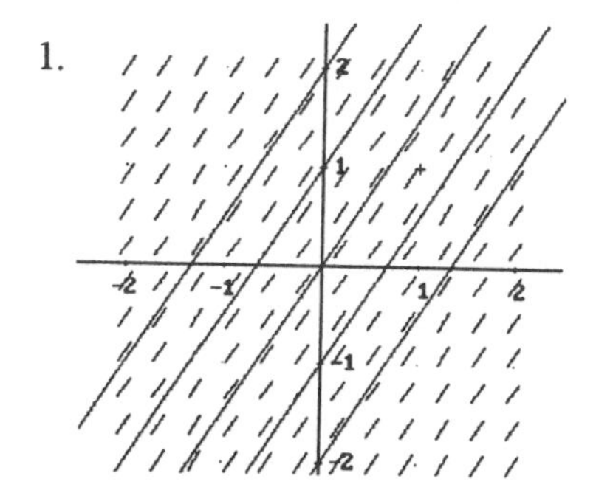

$y = 1.5x + c$

2. (i)

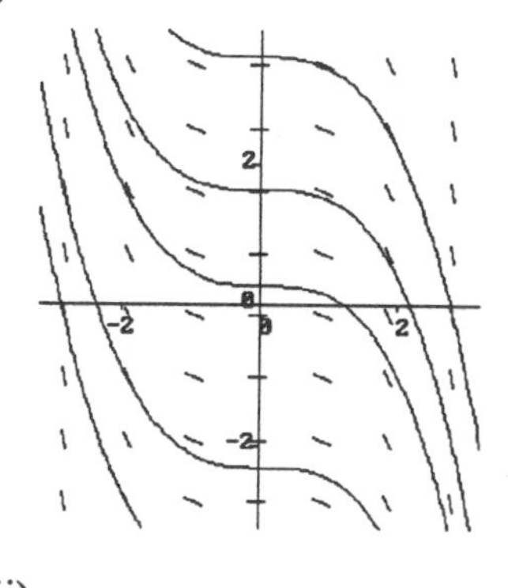

(ii)

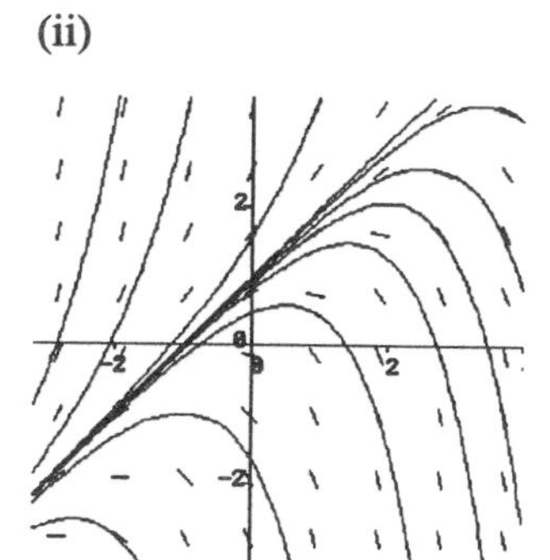

(iii)

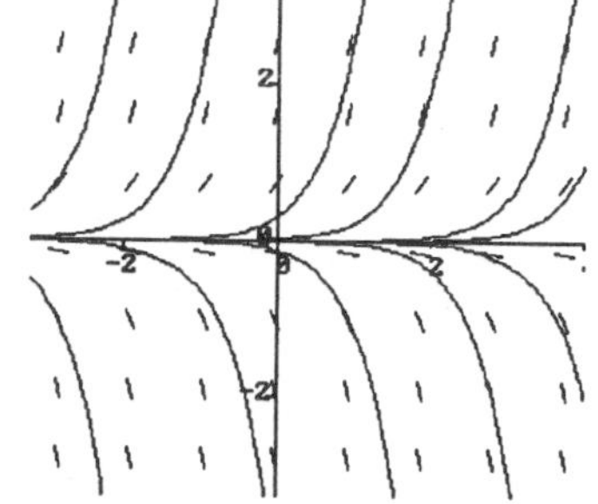

(iv)

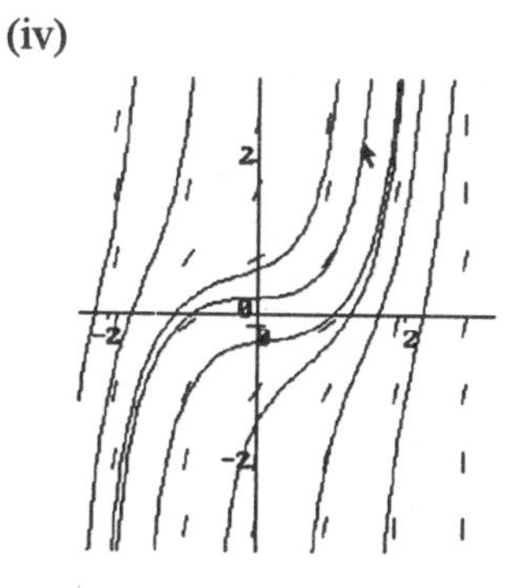

(v)

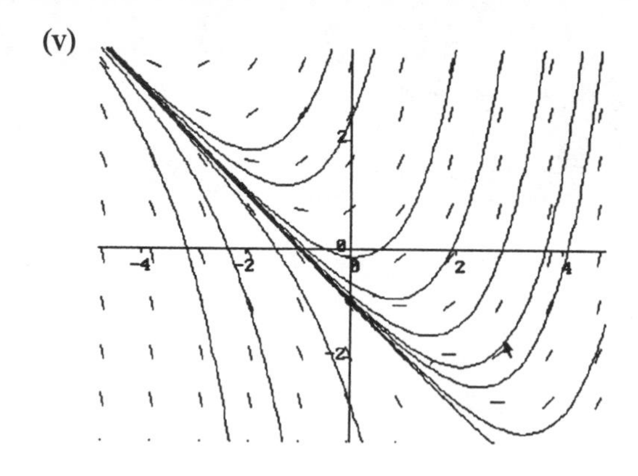

3. (i)

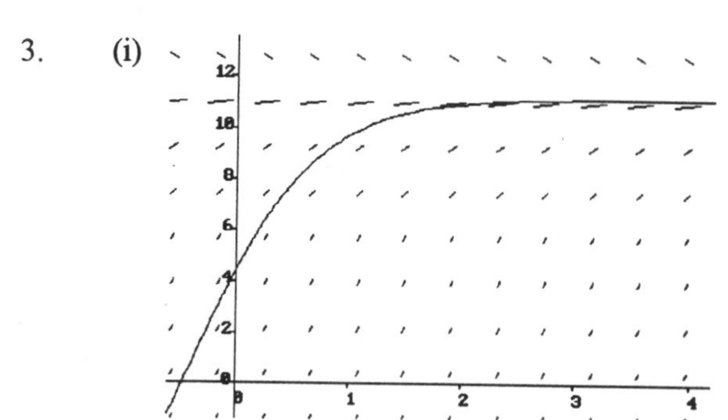

(ii) Parachutist starts with speed 5 ms^{-1} and the speed increases to the terminal velocity 11.18 ms^{-1}.

4. (i)

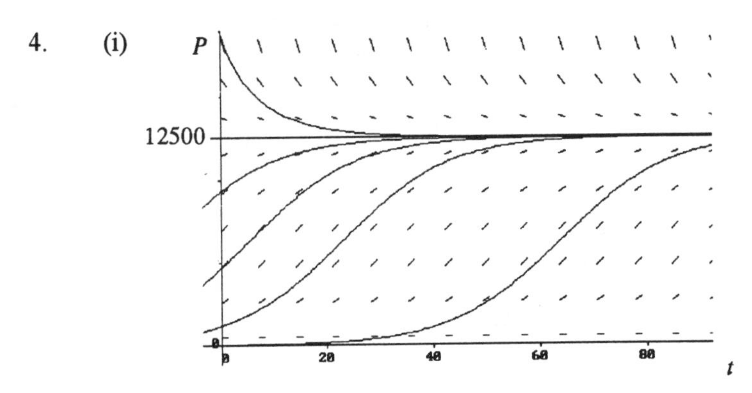

(ii) 12 500 fish.

(iii)

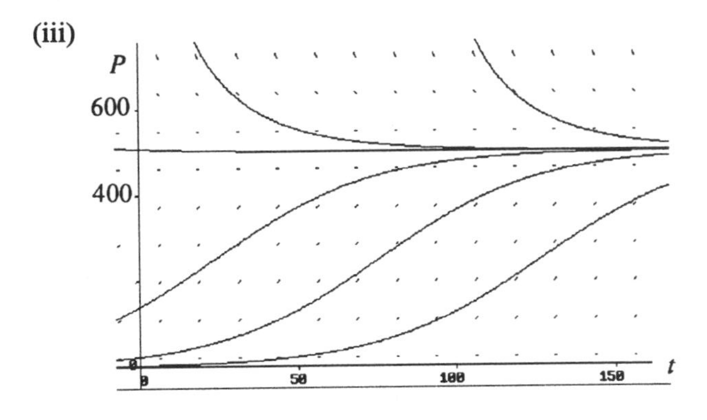

(iv) $0 \leq f < 0.1$

Exercises 1.2

1. (i) integrating factor method
   (ii) variables separable method
   (iii) neither, second order
   (iv) variables separable method
   (v) integrating factor method
   (vi) neither, non-linear in y, cannot separate

2. $x = -\ln\left(1 - \frac{1}{2}t^2\right)$

3. $\theta = 20 + 60e^{-0.05t}$
   13.9 minutes

4. (i) $y = \dfrac{10}{1 - 10\ln x}$

   (ii) $y = \dfrac{6}{3 - 2x^3}$

   (iii) $y^2 = 100 + \frac{2}{3}x^3$

5. $v = \dfrac{40}{41e^{0.2t} - 40}$

6. (i) $y = Ae^{-4x} + \frac{3}{4}x - \frac{3}{16}$
   (ii) $y = -x \ln x + Ax$
   (iii) $y = Ae^{x} + xe^{x}$
   (iv) $y = A\dfrac{1}{x^3} + x^2$
   (v) $x = Ae^{2t} - \frac{5}{2}t - \frac{5}{4}$

Exercises 1.3

(i) $y = Ae^{\frac{1}{5}x} + Be^{x}$

(ii) $x = A\cos\frac{1}{2}t + B\sin\frac{1}{2}t$

(iii) $y = Ae^{x} + Bxe^{x}$

(iv) $x = e^{-\frac{1}{2}t}\left(A\cos\frac{\sqrt{3}}{2}t + B\sin\frac{\sqrt{3}}{2}t\right)$

(v) $y = Ae^{x} + Be^{-3x}$

Exercises 1.4

(i) $\sin x \approx x$

(ii) $x^3 \approx 1 + 3(x-1) + 3(x-1)^2$

(iii) $\frac{1}{x} \approx -\frac{1}{2} - \frac{1}{4}(x+2) - \frac{1}{8}(x+2)^2$

(iv) $\sqrt{x+1} \approx 1 + \frac{1}{2}x - \frac{1}{8}x^2$

(v) $1 + x + \frac{y^2}{x} \approx 6 - 3(x-1) + 4(y-2) + 4(x-1)^2 - 4(x-1)(y-2) + (y-2)^2$

(vi) $2 + xy^2 - \frac{x}{y} \approx 4 + 2(x-1) - (y+1) - (x-1)(y+1) + 2(y+1)^2$

(vii) $\sin(xy) \approx \frac{\pi}{4}x + x\left(y - \frac{\pi}{4}\right)$

(viii) Taylor polynomial does not exist

Exercises 1.5

1. (a) $\mathrm{tr}(\mathbf{A}) = -5$, $\det(\mathbf{A}) = 4$, $\mathbf{A}^{-1} = \frac{1}{4}\begin{bmatrix} -3 & 2 \\ 1 & -2 \end{bmatrix}$

eigenvalues: $-1, -4$; eigenvectors: $[1,-0.5]^{T}$, $[1,1]^{T}$

(b) $\mathrm{tr}(\mathbf{A}) = 2\sqrt{3}$, $\det(\mathbf{A}) = 4$, $\mathbf{A}^{-1} = \frac{1}{4}\begin{bmatrix} \sqrt{3} & -1 \\ 1 & \sqrt{3} \end{bmatrix}$

eigenvalues are complex: $\sqrt{3} + i$, $\sqrt{3} - i$

(c) $\mathrm{tr}(\mathbf{A}) = 5$, $\det(\mathbf{A}) = 6$, $\mathbf{A}^{-1} = \frac{1}{6}\begin{bmatrix} 2 & -9 \\ 0 & 3 \end{bmatrix}$

eigenvalues: 2, 3; eigenvectors: $[9,-1]^{T}$, $[1,0]^{T}$

(d) $\text{tr}(\mathbf{A}) = 0$, $\det(\mathbf{A}) = -1$, $\mathbf{A}^{-1} = \begin{bmatrix} 0 & 1 \\ 1 & 0 \end{bmatrix}$

eigenvalues: 1, –1; eigenvectors: $[1,1]^T$, $[1,-1]^T$

(e) $\text{tr}(\mathbf{A}) = 2$, $\det(\mathbf{A}) = 0$, $\mathbf{A}^{-1}$ does not exist
eigenvalues: 0,2; eigenvectors: $[1,1]^T$, $[1,-1]^T$

(f) $\text{tr}(\mathbf{A}) = 4$, $\det(\mathbf{A}) = 13$, $\mathbf{A}^{-1} = \dfrac{1}{13}\begin{bmatrix} 3 & -5 \\ 2 & 1 \end{bmatrix}$

eigenvalues are complex: $2 + 3i$, $2 - 3i$

2. (i) $[1,-1.5]^T$, $[1,1]^T$
   (ii) $[1,2]^T$, $[1,-2]^T$

Exercises 1.6

1. (i) converges (to 3)
   (ii) diverges
   (iii) diverges
   (iv) converges
   (v) converges

A simple criteria for convergence and a method for finding a limit will be developed in Chapter 4.

Chapter 2

Exercises 2.1

1. (i) and (iv) are autonomous

2. (i) is autonomous

3. $x = x_0 e^{t-t_0}$

4. $x = x_0 e^{\frac{1}{2}(t^2 - t_0^2)}$

5. 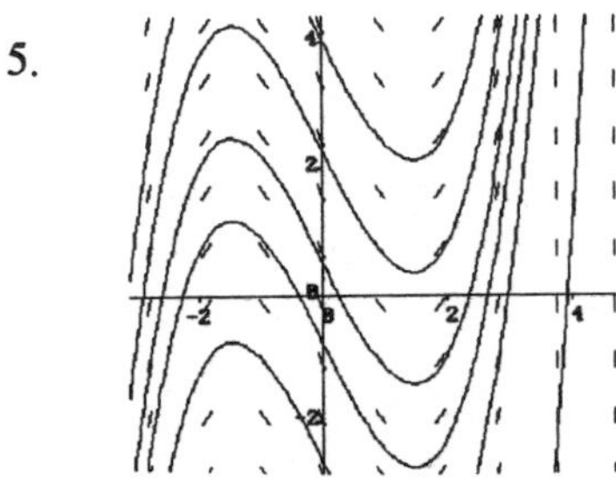

Each solution curve differs by a constant of integration.

6. → → ← ←

$u = 0$ $\quad u = \sqrt{50}$ → u

fixed point

7. Fixed points are $N = a$ and $N = b$

(i) $N < a < b$ → → a b → N

(ii) $a < N < b$ ← ← ← a b → N

(iii) $a < b < N$ → → a b → N

Exercises 2.2

1. Fixed points $x = a$, $x = b$
   Invariant sets $(-\infty,a)$, (a,b), (b,∞)

2. Fixed points $x = n\pi$, $n = 0, \pm1, \pm2, \pm3, \dots$
   Invariant sets $((n-1)\pi, n\pi)$, $n = 0, \pm1, \pm2, \pm3, \dots$

3. $x = x_f$ is the fixed point.
   For $b < 0$, any initial point ($\neq x_f$) moves towards the fixed point.
   For $b > 0$, any initial point ($\neq x_f$) moves away from the fixed point.
   The particular solution is $x = x_f + (x_0 - x_f)e^{bt}$.

4. The fixed points are -1, -0.5, 1 and 2.

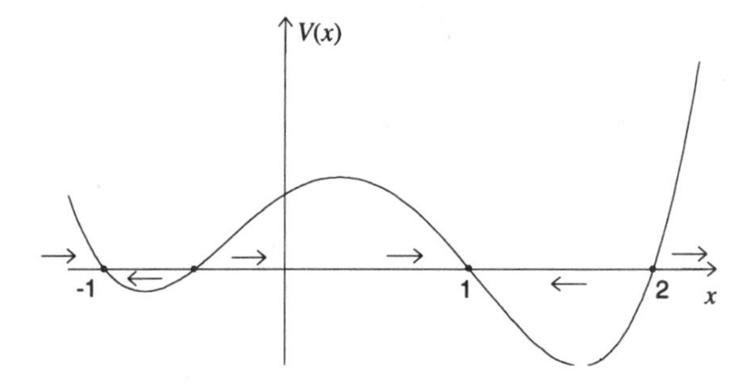

Invariant sets $(-\infty,-1)$, $(-1,-0.5)$, $(-0.5,1)$, $(1,2)$, $(2,\infty)$

Exercises 2.3

2. $x = -1$ is unstable
   $x = 0$ is stable
   $x = 1$ is unstable

3. (i) $x = a$ is stable
   $x = b$ is unstable

   (ii) $x = a$ is unstable
   $x = b$ is stable

4. No, otherwise the system is discontinuous.
   Two essentially different phase diagrams:

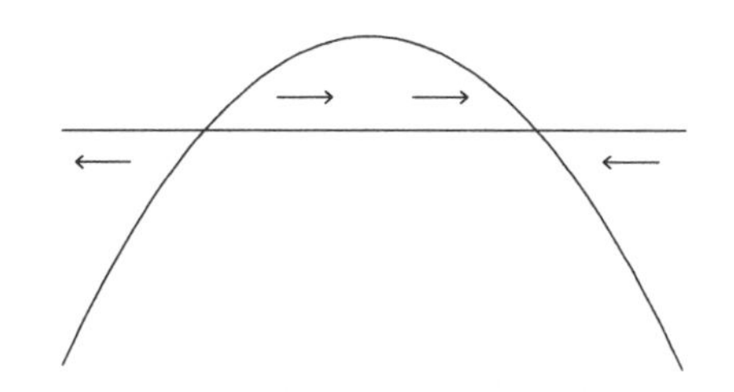

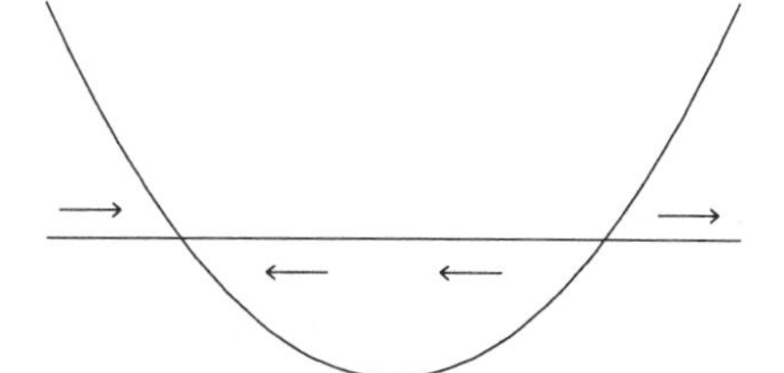

5. $V''(x_f) < 0$: $x_0 < x_f$ $x \to -\infty$
   $x_0 > x_f$ $x \to x_f$

   $V''(x_f) > 0$: $x_0 < x_f$ $x \to x_f$
   $x_0 > x_f$ $x \to \infty$

6. $d > 0$: $t = \dfrac{1}{2d}$ $x \to \pm\infty$ terminating motion
   $d < 0$: $t \to \infty$ $x \to 0$ for all x_0

7. (i) $x = -5$ unstable

   (ii) No fixed points

   (iii) $x = -1$, stable
   $x = 0$, not simple, shunt, semi-stable from above
   $x = 2$, unstable

   (iv) $x = -\frac{1}{2}$, unstable
   $x = 1$, not simple, shunt, semi-stable from below

   (v) $x = 0$, not simple, shunt, semi-stable from below

8. (i) $a > 0$ and $b > \frac{2a}{3}\sqrt{\frac{a}{3}}$ or $a > 0$ and $b < -\frac{2a}{3}\sqrt{\frac{a}{3}}$ or $a < 0$

   (ii) $a > 0;\ -\frac{2a}{3}\sqrt{\frac{a}{3}} < b < \frac{2a}{3}\sqrt{\frac{a}{3}}$

9. $h = 0$, unstable
   $h = 192$, stable

10. (i) m is minimum population level below which the species will become extinct.
    M is the equilibrium population level for $P_0 > m$.

    (ii) (a) $P_0 < m \Rightarrow P \to 0$
    (b) $m < P_0 < M \Rightarrow P \to M$
    (c) $P_0 > M \Rightarrow P \to M$

Exercises 2.4

1. (i) (a) $x_f = 0$, unstable
   $x_f = 1$, not simple, shunt, semi-stable from below
   Invariant sets $(-\infty,0)$, $(0,1)$, $(1,\infty)$

   (b) $t \to 2$, $x \to \infty$ so we have terminating motion

   (c) For $v(x) = x^3 - 2x^2 + x + \varepsilon$
   $\varepsilon > 0$, one unstable fixed point
   $\varepsilon < 0$, three simple fixed points; two are unstable and the other is stable

   (ii) Natural boundary at $x = 0$.
   Imposed natural boundary at $x = 1$.
   Near $x = 0$, system moves away from $x = 0$.
   Near $x = 1$, $x \to 1$ as $t \to \infty$, no terminating motion.

2. (i) No natural boundaries
   $x_f = 0$ is a non-simple shunt

   (ii) $x_0 > 0 \Rightarrow x \to \infty$ as $t \to \infty$

   $x_0 < 0 \Rightarrow x \to 0$ as $t \to -x_0^{\frac{1}{3}}\ (> 0)$

5. For $g(x) < 0$ there is no motion.
   Motion between the zeros of $g(x)$.
   For $g(x) > 0$ there is terminating motion.

Exercises 2.5

1. (i) $P = 0$, unstable, natural boundary
   $P = e^a$, stable

   (ii) $P_0 < e^a$, P increases towards e^a
   $P_0 > e^a$, P decreases towards e^a

   (iii)

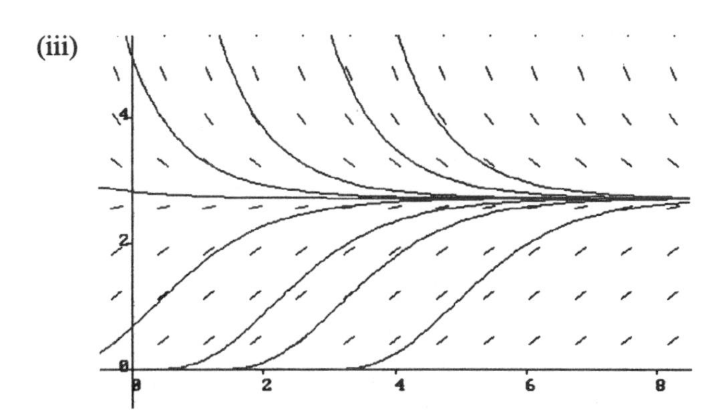

   (iv) As $t \to \infty$, $P \to e^a$

2. $e > g(0)$, the population is doomed to extinction.
   $e < g(0)$, the population will tend towards a stable equilibrium value M given by $g(M) = e$.

3. (i)

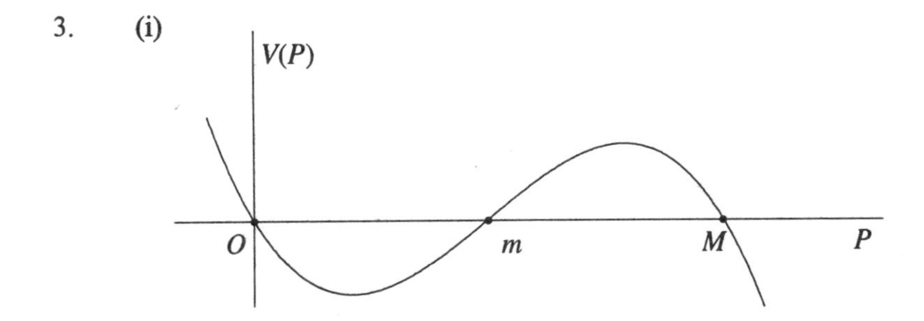

   (ii) (a) $P_c < m$, deer population is doomed to extinction.
   (b) $m < P_c < M$, deer population increases and tends towards M.
   (c) $P_c > M$, deer population decreases and tends towards M.

Further Exercises

1. (i) $x_f = 0$, unstable if $b > 0$, stable if $b < 0$.

   (ii) $x_f = 0$, unstable
   $x_f = 5$, stable

(iii) $x_f = -2$, stable
$x_f = 0$, unstable
$x_f = 2$, stable

(iv) $x_f = -1.5$, stable
$x_f = -1$, unstable
$x_f = \frac{1}{3}$, stable
$x_f = 1$, unstable

(v) $x_f = \dfrac{\pi n}{2}$, $n = 4m + 1$ for integer m, stable fixed points
$x_f = \dfrac{\pi n}{2}$, $n = 4m + 3$ for integer m, unstable fixed points

(vi) $x_f = 0$, unstable natural boundary

2. $\alpha = 0$: $x_f = 0$ is a non-simple unstable fixed point
$\alpha > 0$: $x_f = 0$ is a simple unstable fixed point
$\alpha < 0$: three simple fixed points
$x_f = 0$ is simple and stable
$x_f = \pm\sqrt{-\alpha}$ are simple and unstable

Terminating motion for $\alpha = 0$, terminating time $\dfrac{1}{\sqrt{2}x_0}$.

3. For $v_1(x)$, $x \to \infty$ as $t \to T$. Terminating time is x_0^{-1}.
For $v_2(x)$: $x \to x_f$ as $t \to \infty$

4. Maximum quantity of substance c is $\dfrac{a}{p}$

5. (i) $v \to \sqrt{\dfrac{9.8m}{k}}$ as $t \to \infty$ for all initial speeds $v(0)$

(ii) $v(t) = v_c \dfrac{((v_c + v_0)e^{2gt/v_c} + v_0 + v_c)}{(v_c + v_0)e^{2gt/v_c} + v_c - v_0}$ where $v_0 = v(0)$ and $v_c = \sqrt{\dfrac{9.8m}{k}}$.
As $t \to \infty$, $v \to v_c$.

6. (i) Let h be the depth (metres) at time t (hours)

$$\frac{dh}{dt} = -\frac{1}{35}(1 + 2h)$$

(ii) Natural boundary at $h = 0$
No fixed points

(iii) Depth of water reduces to zero.

7. (i) Let m be the mass (grams) deposited at time t (seconds)

$$\frac{dm}{dt} = km(20 - m)$$

(ii) Natural boundary at $m = 0$ and $m = 20$
Fixed points at $m = 0$, $m = 20$

(iii) $0 < m_0 < 20$, m increases to the equilibrium value $m = 20$

8. (i) Natural boundary and fixed point at $x = 2$
(ii) For $x_0 < 2$, x tends towards $x = 2$
(iii) Terminating time $T = \sqrt{2}$

Chapter 3

Exercises 3.1

1. (i) is autonomous.

2. (i) $\dfrac{dx_1}{dt} = x_2$ and $\dfrac{dx_2}{dt} = -\lambda \sin x_1$

(ii) $\dfrac{dx_1}{dt} = x_2$ and $\dfrac{dx_2}{dt} = -b - 2ax_2$

(iii) $\dfrac{dx_1}{dt} = x_2$ and $\dfrac{dx_2}{dt} = -\lambda x_1 + k \sin \omega t$

3. $\mathbf{v} = (x_2, -g)$

4. $\mathbf{v} = (10x_2, 0.98 - 5x_2^2)$

5. $\mathbf{v} = (2x_1 - \cos 3x_2, 1)$
$\mathbf{v} = (v(x_1, x_2), 1)$

Exercises 3.2

1. (i) $x_1 = e^t$ and $x_2 = 2e^t$

(ii) $$x_1 = \frac{(3+\sqrt{2})}{2\sqrt{2}} e^{\sqrt{2}t} + \frac{(\sqrt{2}-3)}{2\sqrt{2}} e^{-\sqrt{2}t}$$
$$x_2 = \frac{(2\sqrt{2}-1)}{2\sqrt{2}} e^{\sqrt{2}t} + \frac{(2\sqrt{2}+1)}{2\sqrt{2}} e^{-\sqrt{2}t}$$

(iii) $$x_1 = \frac{2}{\sqrt{6}} e^t \sin\sqrt{6}t + 1$$
$$x_2 = e^t \cos\sqrt{6}t + 1$$

(iv) $$x_1 = \frac{3}{2} e^{5t} - \frac{1}{2} e^{-t}$$
$$x_2 = \frac{3}{2} e^{5t} + \frac{1}{2} e^{-t}$$

2. $a > b + c$
limiting value is $(a - b - c)/(a - c)$

3. population of predators and preys increase indefinitely

Exercises 3.3

1. (i) *A* strongly unstable

   (ii) *B* strongly stable
   C unstable

   (iii) *A* unstable
   B strongly stable

   (iv) *C* unstable
   D stable centre

2. (i) (0,0) unstable
   (ii) (–2,1) strongly unstable
   (iii) (0,0) strongly unstable
   (–1,1) unstable
   (–1,–1) unstable
   (–2,0) strongly stable
   (iv) (4,2) strongly unstable
   (–2,–1) unstable
   (v) (2,2) strongly stable
   (–1,–1) strongly unstable
   (–2,2) unstable
   (1,–1) unstable

5. $$\frac{dx_1}{dt} = x_2 \qquad \frac{dx_2}{dt} = 1 - x_1 - x_2^2$$
Fixed point at (1,0)

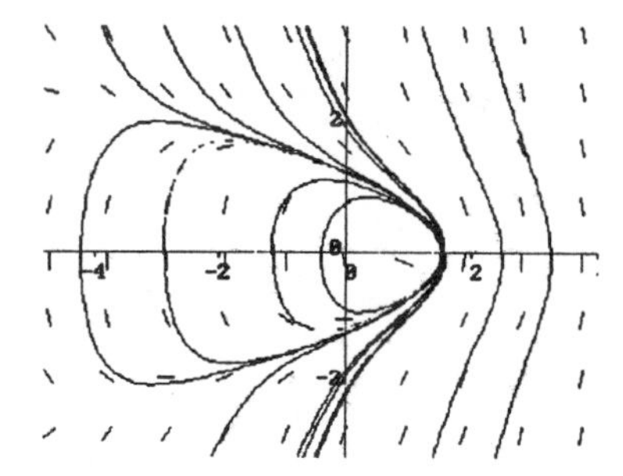

strongly unstable fixed point

6. (i) $\dfrac{dx_1}{dt} = x_2 \qquad \dfrac{dx_2}{dt} = -x_1 + x_1^3 - x_2$

Fixed points at (0,0), (–1,0), (1,0)

(ii) $\dfrac{dx_1}{dt} = x_2 - x_1 \qquad \dfrac{dx_2}{dt} = x_1^3 - x_1$

Fixed points at (0,0), (–1,–1), (1,1)

The direction field for each system is

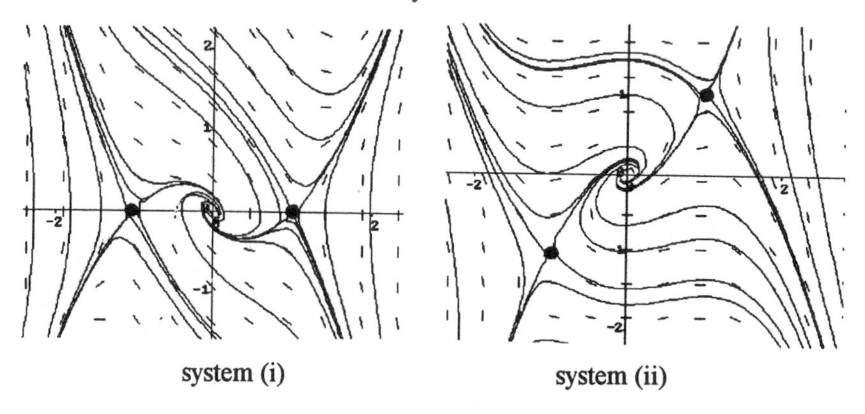

system (i) system (ii)

Exercises 3.4

2. (i) Unstable node
   (ii) Saddle
   (iii) Unstable star
   (iv) Saddle
   (v) Stable star

3. (i) $x_1 = c_1 e^{2t} + c_2 e^{3t}$

   $x_2 = c_2 e^{3t}$

   (ii) $x_1 = 2c_1 e^{5t} + c_2 e^{-t}$

   $x_2 = c_1 e^{5t} - c_2 e^{-t}$

(iii) $x_1 = c_1 \cos 2t + c_2 \sin 2t$

$x_2 = \frac{1}{5}(c_1 - 2c_2)\cos 2t + \frac{1}{5}(c_2 + 2c_1)\sin 2t$

(iv) $x_1 = c_1 e^{\sqrt{5}t} + c_2 e^{-\sqrt{5}t}$

$x_2 = (\sqrt{5} - 2)c_1 e^{\sqrt{5}t} - (\sqrt{5} + 2)c_2 e^{-\sqrt{5}t}$

4. (i) $\mathbf{P} = \begin{bmatrix} 1 & 1 \\ -1 & 0 \end{bmatrix}$ $\mathbf{M} = \begin{bmatrix} 3 & 1 \\ 0 & 3 \end{bmatrix}$

(ii) $\mathbf{P} = \begin{bmatrix} 1 & 1 \\ -1.5 & 1 \end{bmatrix}$ $\mathbf{M} = \begin{bmatrix} -1 & 0 \\ 0 & 4 \end{bmatrix}$

(iii) $\mathbf{P} = \begin{bmatrix} 1 & 1 \\ 0 & 1 \end{bmatrix}$ $\mathbf{M} = \begin{bmatrix} 2 & 1 \\ 0 & 2 \end{bmatrix}$

(iv) $\mathbf{P} = \begin{bmatrix} \sqrt{2} & \sqrt{2} \\ 1 & -1 \end{bmatrix}$ $\mathbf{M} = \begin{bmatrix} \sqrt{2}+1 & 0 \\ 0 & 1-\sqrt{2} \end{bmatrix}$

(v) $\mathbf{P} = \begin{bmatrix} -1 & -\sqrt{3} \\ 2 & 0 \end{bmatrix}$ $\mathbf{M} = \begin{bmatrix} 2 & -\sqrt{3} \\ \sqrt{3} & 2 \end{bmatrix}$

(vi) $\mathbf{P} = \begin{bmatrix} 1 & -0.5 \\ -2 & 0 \end{bmatrix}$ $\mathbf{M} = \begin{bmatrix} 2 & 1 \\ 0 & 2 \end{bmatrix}$

5. (i) $(-1,-1)$ unstable node

(ii) $(\frac{1}{2},-\frac{1}{2})$ stable spiral

(iii) $(-\frac{7}{4},\frac{17}{4})$ unstable star

(iv) $(4,-3)$ unstable saddle

(v) $(1,\frac{1}{2})$ unstable saddle

Exercises 3.5

1. (i) (4,4) unstable spiral

(–2,–2) stable spiral

(ii) (–1,–1) unstable spiral

(2,2) stable node

(1,–1) saddle

(–2,2) saddle

(iii) (–1,1) unstable spiral

(iv) (1,1) stable node

(2,2) saddle

(1,–1) saddle

(2,–2) unstable spiral

(v) (0,0) unstable node

(2,2) stable star

(vi) (1,0) unstable node
(–1,0) saddle
$(e^{-0.5},-1)$ saddle
$(-e^{-0.5},-1)$ stable node

2. (i) (0,0) saddle
(2,0) unstable spiral

(ii) Linear system has a centre at (–5,0) so the non-linear system has a centre or a spiral.

3.

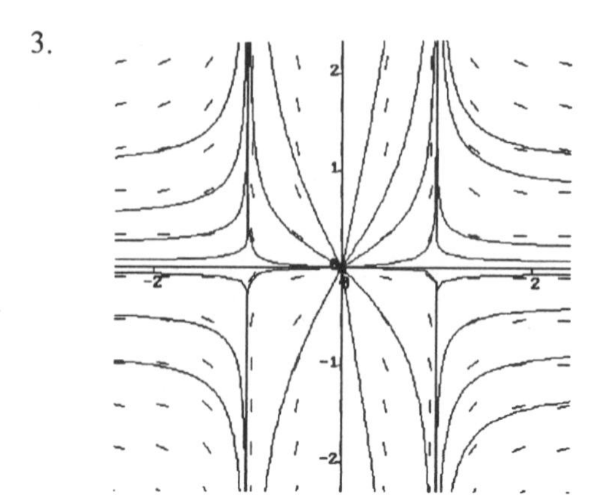

4. (i) Stable coexistence is possible with $x = \frac{24}{7}$ and $y = \frac{40}{7}$.
(ii) Each fixed point is unstable; both species will become extinct.

6. (0,0) is a centre for linearized system.
Non-linear system could have a centre or a spiral.

7. (i) Limit cycle $r = 1$, attracting for $r < 1$ and repelling for $r > 1$.
(ii) Limit cycle $r = 2$, attracting for $r < 2$
r = constant outside $r = 2$.

Exercises 3.6

1.
$$\frac{dp_1}{dt} = I - \frac{ap_1}{v_1} - \frac{bp_1}{v_1}$$
$$\frac{dp_2}{dt} = \frac{ap_1}{v_1} - \frac{cp_2}{v_2}$$
$$\frac{dp_3}{dt} = \frac{bp_1}{v_1} + \frac{cp_2}{v_2} - \frac{dp_3}{v_3}$$

where $a + b = d$ and $b + c = d$

3. $H = 0$, competitive exclusion
   $H = \frac{5}{4}$, unstable coexistence
   $H = \frac{25}{4}$, extinction

Further Exercises

2. $ad - b^2 < 0$, saddle
   $ad - b^2 > 0$, $a + d > 0$, unstable node
   $ad - b^2 > 0$, $a + d < 0$, stable node
   $b = 0$ and $a = d$, star, stable if $a < 0$, unstable if $a > 0$

3. $a - d > 2b$, unstable node
   $a - d < 2b$, unstable spiral
   $a - d = 2b$ $b \neq 0$, unstable node; $b = 0$, unstable star

4. (i) Non-simple fixed point
   (ii) Stable node
   (iii) Unstable spiral
   (iv) Unstable node

5. All except a centre

6. (i) (1,1) saddle
   (ii) (4,4) unstable spiral
   (–1,–1) stable spiral
   (iii) (0,2) centre or spiral
   (0,–2) centre or spiral
   (1,0) saddle
   (–1,0) saddle
   (iv) (0,1) centre or spiral
   (0,–1) saddle
   (v) (0,0) not simple
   (–2,2) saddle

8. (0,0) saddle
   (1,1) unstable spiral
   (–1,0) stable node

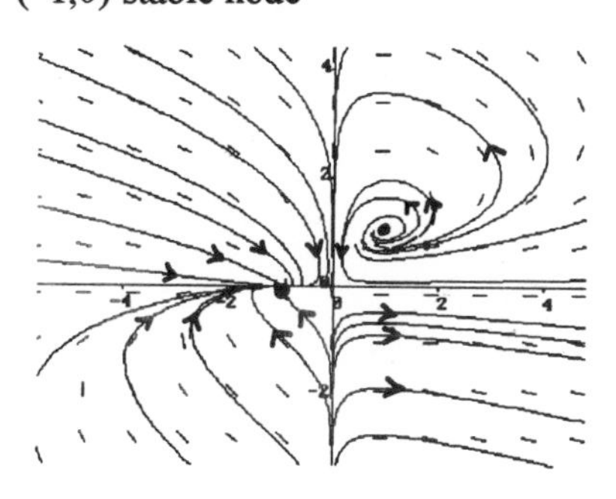

9. (i) $\frac{dx_1}{dt} = x_2, \quad \frac{dx_2}{dt} = -x_1 - x_1^3$
   (ii) (0,0) which is a centre
   (iii) Spiral

10. $x_1 = \theta, \quad x_2 = \dot{\theta}$
    $(\pm n\pi, 0)$ n even, stable node $\alpha^2 > 4w^2$; stable spiral points $\alpha^2 < 4w^2$
    $(\pm n\pi, 0)$ n odd, saddle points
    $\alpha = 0$, centre for n even, saddle for n odd
    Pendulum rotates with damping gradually reducing motion until it spirals into a fixed point.

11. (i) (25.82, 774.60) unstable node
    P, Q tend to zero

13. (0,0) unstable node
    (–40,0) not realistic
    (0,1200) stable node
    (10,1000) saddle

Chapter 4

Exercises 4.1

1. £1448.67

2. We assume that building societies adopt the following method:

   - at the beginning of each year the interest on the outstanding loan is added to the outstanding amount;
   - the 12 monthly repayments M are then subtracted to give the amount owed at the end of the year;
   - this process is repeated 25 times and M chosen so that the amount owed is then zero.

   (i) $(1.095 \times 30\ 000) - 12M$
   (ii) $x_{n+1} = 1.095x_n - 12M$
   (iii) £264.90
   (iv) £22 200

3. (i) $P_{n+1} = 1.25P_n, P_0 = 5000$
   (ii) 15 258
   (iii) During 1996

4. $P_{n+1} - P_n = (0.7 - 3 \times 10^{-5} P_n)P_n$
   Stable equilibrium is 2.33×10^4 birds.
   With shooting: $P_{n+1} - P_n = (0.5 - 3 \times 10^{-5} P_n)P_n$.
   New stable equilibrium is 1.67×10^4 birds.
   The population of birds is within 1% of the equilibrium population after 7 years.

5. $P_{n+1} - P_n = (0.6 - 3 \times 10^{-4} P_n)P_n$
   Stable equilibrium is 2000 fish.
   With restocking: $P_{n+1} - P_n = (0.8 - 3 \times 10^{-4} P_n)P_n$
   New stable equilibrium is 2666 fish.
   The stock level is within 5% of this equilibrium population after 3 years.

6. (i) $x_0 = 1.0$ $x_5 = 1563$
   (ii) $x_0 = 1.1$ $x_5 = 1875.5$
   (iii) $x_0 = 0.9$ $x_5 = 1250.5$
   (iv) $x_0 = 0.5$ $x_5 = 0.5$
   (v) $x_0 = 0.51$ $x_5 = 31.75$
   (vi) $x_0 = 0.49$ $x_5 = -30.75$

   A small change in x_0 leads to a large change in x_5 so that the recurrence relation is said to be ill-conditioned.

7. (i) converges to 3.8063 in 4 terms,
   (ii), (iv) converge in greater than 20 terms,
   (iii) is a periodic cycle.

Exercises 4.2

1. (i) $F(x) = x + 3$, $F^2(x) = x + 6$, $F^3(x) = x + 9$
   No fixed points

   (ii) $F(x) = x^2 - x + 2$, $F^2(x) = x^4 - 2x^3 + 4x^2 - 3x + 4$
   $F^3(x) = x^8 - 4x^7 + 12x^6 - 22x^5 + 35x^4 + 37x^2 - 21x + 14$
   No fixed points

   (iii) $F(x) = x^3 - 3x$, $F^2(x) = x^9 - 9x^7 + 27x^5 - 30x^3 + 9x$
   $= x(x^2 - 3)(x^6 - 6x^4 + 9x^2 - 3)$
   Fixed points: 0, 2, –2

   (iv) $F(x) = \cos x$, $F^2(x) = \cos(\cos(x))$
   Fixed point: 0.7391 (to 4 dp)
   $F(x) = |x|$, $F^2(x) = |x|$
   Fixed points: all $x \geq 0$

2. (i) $\frac{1}{2}, \frac{1}{4}, \frac{1}{16}, \frac{1}{256}, \frac{1}{65536}$

   or 0.5, 0.25, 0.0625, 0.003 906 25, 0.000 015 258 789 0625

   (ii) 0, 1, 2, 5, 26

   (iii) 1, 0.841 47, 0.745 623, 0.678 430, 0.627 571

   (iv) $\frac{1}{4}, \frac{1}{2}$, 0, 0, 0

3. Period-2 orbit is $\sqrt{2}, -\sqrt{2}$

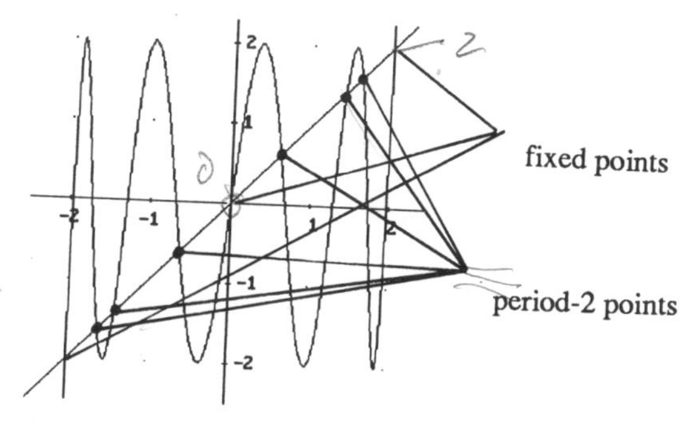

4. (i) $x_f = 2$

   (ii) $F^2(x) = 0.25x + 1.5$
   $F^3(x) = 0.125x + 1.75$
   $F^4(x) = 0.0625x + 1.875$
   $F^n(x) = 0.5^n x + 2(1 - 0.5^n)$

   (iii) Only solution of $F^n(x) = x$ is $x = 2$.

5. (i) $1+\sqrt{3}, 1-\sqrt{3}$

   (iii)

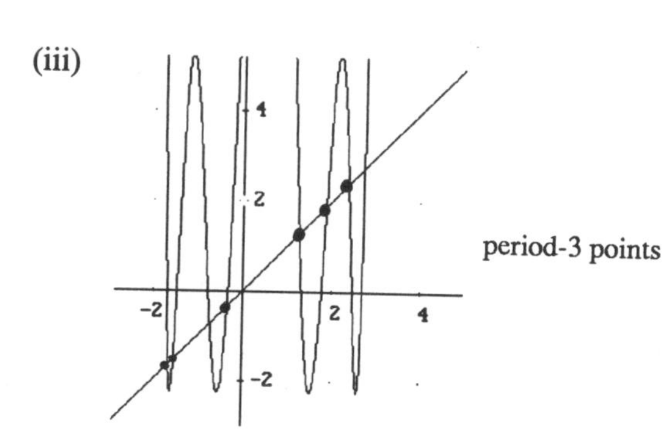

Period-3 points: –1.65, 2.38, 1.28; –0.35, –1.53, 1.88

6. (i) Fixed points 0, $\frac{2}{3}$

(ii)

$$T^2(x) = \begin{cases} 4x & 0 \le x \le \frac{1}{4} \\ 2-4x & \frac{1}{4} \le x \le \frac{1}{2} \\ -2+4x & \frac{1}{2} \le x \le \frac{3}{4} \\ 4-4x & \frac{3}{4} \le x \le 1 \end{cases}$$

$$T^3(x) = \begin{cases} 8x & 0 \le x \le \frac{1}{8} \\ 2-8x & \frac{1}{8} \le x \le \frac{1}{4} \\ -2+8x & \frac{1}{4} \le x \le \frac{3}{8} \\ 4-8x & \frac{3}{8} \le x \le \frac{1}{2} \\ -4+8x & \frac{1}{2} \le x \le \frac{5}{8} \\ 6-8x & \frac{5}{8} \le x \le \frac{3}{4} \\ -6+8x & \frac{3}{4} \le x \le \frac{7}{8} \\ 8-8x & \frac{7}{8} \le x \le 1 \end{cases}$$

(iii)

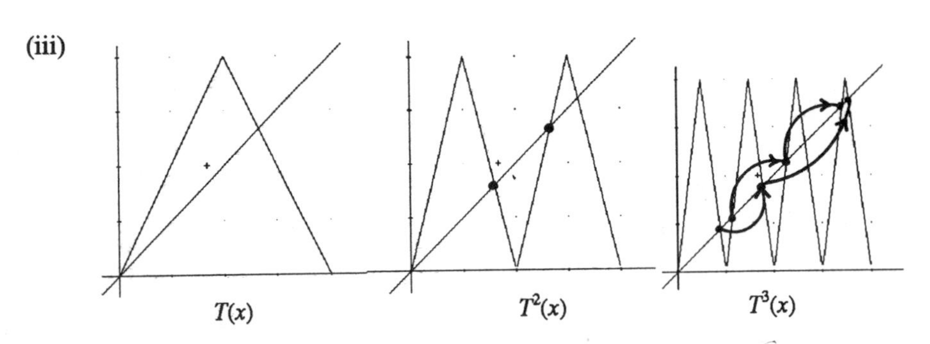

Period-2 orbits: $\frac{2}{5}, \frac{4}{5}$

Period-3 orbits: $\frac{2}{9}, \frac{4}{9}, \frac{8}{9}$; $\frac{2}{7}, \frac{4}{7}, \frac{6}{7}$

7. (ii) $\frac{1}{3}, \frac{2}{3}$

(iii) $\frac{1}{5}, \frac{2}{5}, \frac{4}{5}, \frac{3}{5}$

(iv) $\frac{1}{9}, \frac{2}{9}, \frac{4}{9}, \frac{8}{9}, \frac{7}{9}, \frac{5}{9}$

(v)

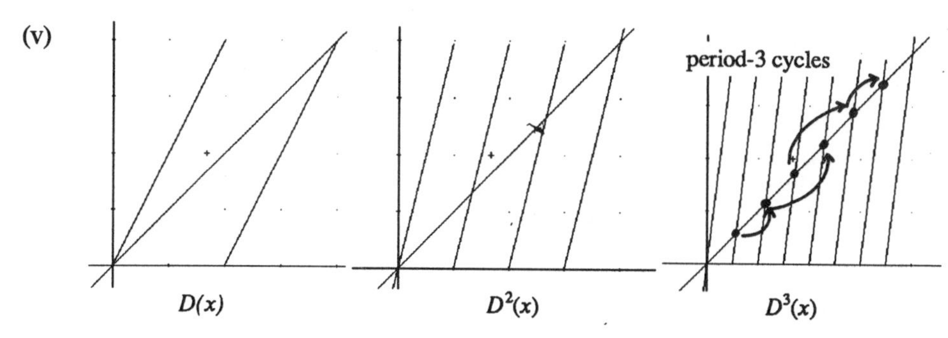

Period-3 cycles: $\frac{1}{7}, \frac{2}{7}, \frac{4}{7}$; $\frac{3}{7}, \frac{6}{7}, \frac{5}{7}$

Exercises 4.3

1. $F^2(x) = a(b+1) + b^2x$
   $F^3(x) = a(b^2 + b + 1) + b^3x$

2. $F^n(x) = a(1 + b + b^2 + \ldots + b^{n-1}) + b^n x = \dfrac{a(1-b^n)}{(1-b)} + b^n x$

3. (i) $F^2(x) = |x|$
   $F^3(x) = |x|$
   (ii) Any points $x \geq 0$ are fixed points.
   (iii) No periodic orbits.

Exercises 4.4

1. (i) Unstable fixed point at $x = \frac{1}{4}$

   (ii) Fixed point at $x = 0$, attractor for $x_0 < 0$ and repellor for $x_0 > 0$, semi-stable from below

   (iii) Fixed point at $x = 0$, repellor for $x_0 < 0$ and attractor for $x_0 > 0$, semi-stable from above

   (iv) Fixed point at $x = 0$ which is an attractor for all x_0

   (v) Fixed point at $x = 0$ which is an attractor for all x_0

2. (i) Stable fixed point at $x = 2$

   (ii) Stable fixed point at $x = 1$
   Fixed point at $x = 0$ which is a repellor for $x_0 > 0$.
   For $x_0 < 0$ map does not exist so $x = 0$ is a natural boundary.

(iii) Fixed points at $x = \pm 1$

Every seed $x_0 \neq 0$ lies on a period-2 orbit $x_0, \dfrac{1}{x_0}$

(iv) No fixed points

Orbit diverges for all x_0

(v) Stable fixed point at $x = 0$

Unstable fixed point at $x = -1$

(vi) Unstable fixed point at $x = 0$

Stable fixed points at $x = -0.6$ and $x = 0.6$

Period-2 orbit $0.2\sqrt{59}, \ -0.2\sqrt{59}$

3. (i) Unstable fixed point at $0.5 + 0.3\sqrt{15}$ (1.661 89)

Stable fixed point at $0.5 - 0.3\sqrt{15}$ (–0.661 89)

(ii) $x^4 - 2.2x^2 + 0.11$

(iii) $0.1\sqrt{35} - 0.5, \ -0.1\sqrt{35} - 0.5$

(iv)

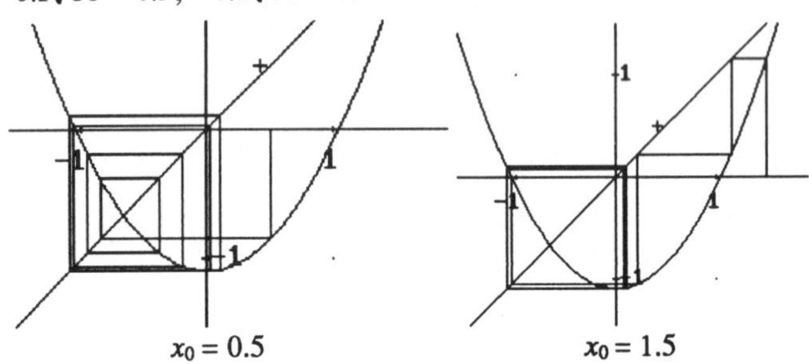

4. (i) Fixed point 0

Eventually fixed point $x_0 = 1$

(ii) Fixed points $0, \frac{2}{3}$

Eventually fixed points $x_0 = 1, \ x_0 = \frac{1}{3}, \ x_0 = \frac{1}{2} + \frac{1}{6}\sqrt{5}, \ x_0 = \frac{1}{2} - \frac{1}{6}\sqrt{5}$

(iii) Fixed points $-1, 2, \frac{1}{2}(\sqrt{5} - 1), -\frac{1}{2}(\sqrt{5} + 1)$

Many eventually fixed points found by solving $F(x_0) = x_f$, for example with $x_f = -1, x_0 = 1, \sqrt{3}, -\sqrt{3}, \pm\frac{1}{2}(\sqrt{6} \pm \sqrt{2})$

(iv) Fixed points for all $x \geq 0$

All points $x_0 < 0$ are eventually fixed points

5. (i) Fixed points –2, 0, 2

(ii) Period-2 orbits: $\sqrt{2}, -\sqrt{2}$; $\frac{1}{2}(1+\sqrt{5}), \frac{1}{2}(1-\sqrt{5})$; $\frac{1}{2}(\sqrt{5}-1)$, $-\frac{1}{2}(\sqrt{5}+1)$

(iii) $\sqrt{2}, -\sqrt{2}$ repelling
$\frac{1}{2}(1+\sqrt{5}), \frac{1}{2}(1-\sqrt{5})$ repelling
$\frac{1}{2}(\sqrt{5}-1), -\frac{1}{2}(1+\sqrt{5})$ repelling

(iv)

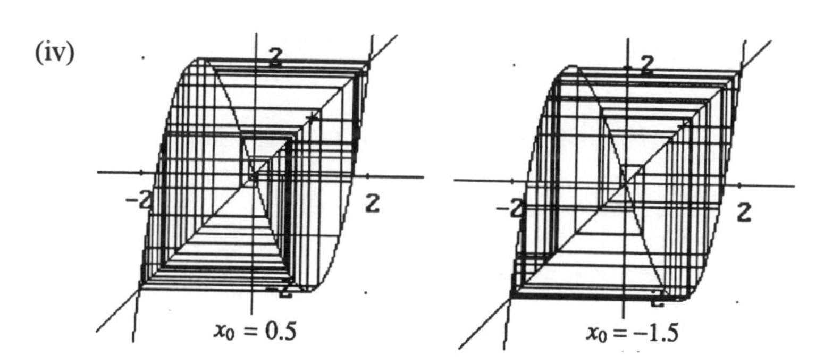

6. $a > 0$, stable fixed point at $\frac{1}{2}(1+\sqrt{1+4a})$ for $x_0 \geq -4a$
$-\frac{1}{4} \leq a \leq 0$ and $x_0 \geq \frac{1}{2}(1-\sqrt{1+4a})$, sequence tends to $\frac{1}{2}(1+\sqrt{1+4a})$
Fixed point $\frac{1}{2}(1-\sqrt{1+4a})$ is unstable

7. For $a > e^{1/e}$, no fixed points, sequence diverges
For $a = e^{1/e}$, fixed point $x_f = e$, semi-stable from below
For $a < e^{1/e}$, two fixed points which are solutions to $a^x = x$;
smaller root is stable; larger root is unstable

8. (i) Fixed points $0, \sqrt{a-1}, -\sqrt{a-1}$

(iii) $\frac{1}{2}(\sqrt{a+2}-\sqrt{a-2}), \frac{1}{2}(\sqrt{a+2}+\sqrt{a-2})$ attracting period-2 cycle for $2 < a < \sqrt{5}$
$\frac{1}{2}(\sqrt{a-2}-\sqrt{a+2}), -\frac{1}{2}(\sqrt{a+2}+\sqrt{a-2})$ attracting period-2 cycle for $2 < a < \sqrt{5}$

Exercises 4.5

1. (i) $x_f = 0$, stable if $k < 1$
$x_f = 1 - \dfrac{1}{k}$, stable if $1 < k < 3$

(ii) $k = 3$

(iii) Behaviour for $0 \leq k \leq 4$ is similar to $x_n^2 - c$ for $0 \leq c \leq 2$

Further Exercises

1. (i) $x_f = 0, x_f = 1$

   (ii) Period-2 cycle: $\dfrac{41+\sqrt{41}}{42}, \dfrac{41-\sqrt{41}}{42}$ (or 1.128 65, 0.823 735 to 6 sf)

2. $c = 1.754\ 88$

3. (i) $\dfrac{43+\sqrt{129}}{66}, \dfrac{43-\sqrt{129}}{66}$ (0.823 603, 0.479 427 to 6 sf)

4. 0.535 947 556, 1.157 716 989, 0.701 237 895, 1.224 996 169

5. (i) $x_f = 0$, stable for $c < 1$ and unstable for $c \geq 1$

   $x_f = \dfrac{c}{1+c}$ for $c \geq 1$, unstable

   For $c < 1.0$ sequence converges to $x = 0$.

   For $c = 1$, $x_n = 0.5$ for all n.

   For $1 < c < 2$ the sequence is trapped by the interval $c(1 - 0.5c) \leq x \leq 0.5c$.

   (ii)

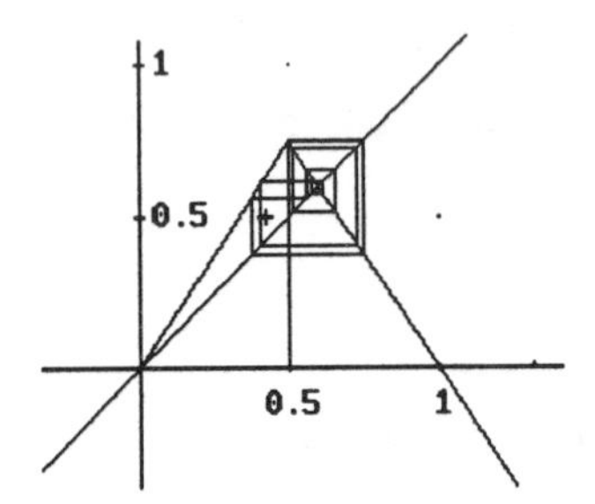

cobweb diagram for $c = 1.5$

6. (ii) $1 < c < 3$

   (iv) $0 < c < 1$

Bibliography

There are many books which develop the introductory ideas of non-linear dynamical systems further and the following list is just a few.

Arrowsmith, D. K., Place, C. M. 1992: *Dynamical systems, differential equations, maps and chaotic behaviour*. London: Chapman and Hall.

Barnsley, M., 1988: *Fractals everywhere*. Boston: Academic Press.

Devaney, R. L., 1992: *A first course in chaotic dynamical systems theory and experiment*. Wokingham: Addison Wesley.

Gleick, J., 1987: *Chaos, making a new science*. London: Abacus.

Lorenz, E., 1993: *The essence of chaos*. London: UCL Press.

Sandefur, J. T., 1990: *Discrete dynamical systems, theory and applications*. Oxford: Clarendon Press.

Index

UNIVERSITY OF GLAMORGAN
PRIFYSGOL MORGANNWG